AF786732

Distinguished Dissertations

Springer-Verlag London Ltd.

Other titles published in this Series:

Search and Planning Under Incomplete Information: A Study Using Bridge Card Play
Ian Frank

Hardware Evolution: Automatic Design of Electronic Circuits in Reconfigurable Hardware by Artificial Evolution
Adrian Thompson

User-Developer Cooperation in Software Development
Eamonn O'Neill

A Combination of Geometry Theorem Proving and Nonstandard Analysis, with Application to Newton's Principia
Jacques Fleuriot

Accurate Visual Metrology from Single and Multiple Uncalibrated Images
Antonio Criminisi

Inheritance Relationships for Disciplined Software Construction
Tracy A. Gardner

Asynchronous System-on-Chip Interconnect
John Bainbridge

Stochastic Algorithms for Visual Tracking
John MacCormick

Automated Theory Formation in Pure Mathematics
Simon Colton

Dynamic Flexible Constraint Satisfaction and its Application to AI Planning
Ian Miguel

Strength or Accuracy: Credit Assignment in Learning Classifier Systems
Tim Kovacs

David Capel

Image Mosaicing and Super-resolution

Springer

Dr David Capel

Series Editor
Professor C.J. van Rijsbergen
Department of Computing Science, University of Glasgow, G12 8RZ, UK

British Library Cataloguing in Publication Data
A catalogue record for this book is available from the British Library

Library of Congress Cataloging-in-Publication Data
Capel, David, 1974-
Image mosaicing and super-resolution / David Capel.
p. cm. – (Distinguished dissertations)
Includes index.
ISBN 978-1-4471-1049-1 ISBN 978-0-85729-384-8 (eBook)
DOI 10.1007/978-0-85729-384-8
1. Image processing—Digital techniques—Mathematics. 2. Resolution (Optics) 3. Matching theory. I. Title. II. CPHC/BCS distinguished dissertations.

TA1637.C475 2003
621.36'7—dc22

2003061882

Apart from any fair dealing for the purposes of research or private study, or criticism or review, as permitted under the Copyright, Designs and Patents Act 1988, this publication may only be reproduced, stored or transmitted, in any form or by any means, with the prior permission in writing of the publishers, or in the case of reprographic reproduction in accordance with the terms of licences issued by the Copyright Licensing Agency. Enquiries concerning reproduction outside those terms should be sent to the publishers.

Distinguished Dissertations ISSN 1439-9768

ISBN 978-1-4471-1049-1

springeronline.com

© Springer-Verlag London 2004
Originally published by Springer-Verlag London Berlin Heidelberg in 2004
Softcover reprint of the hardcover 1st edition 2004

The use of registered names, trademarks etc. in this publication does not imply, even in the absence of a specific statement, that such names are exempt from the relevant laws and regulations and therefore free for general use.

The publisher makes no representation, express or implied, with regard to the accuracy of the information contained in this book and cannot accept any legal responsibility or liability for any errors or omissions that may be made.

Typesetting: Electronic text files prepared by author

34/3830-543210 Printed on acid-free paper SPIN 10945223

Image Mosaicing and Super-resolution

Abstract

This book investigates the problem of how information contained in multiple, overlapping images of the same scene may be combined to produce images of superior quality. This area, generically titled *frame fusion*, offers the possibility of reducing noise, extending the field of view, removal of moving objects, removing blur, increasing spatial resolution and improving dynamic range. As such, this research has many applications in fields as diverse as forensic image restoration, computer generated special effects, video image compression, and digital video editing.

An essential enabling step prior to performing frame fusion is *image registration*, by which an accurate estimate of the point-to-point mapping between views is computed. A robust and efficient algorithm is described to automatically register multiple images using only information contained within the images themselves. The accuracy of this method, and the statistical assumptions upon which it relies, are investigated empirically.

Two forms of frame-fusion are investigated. The first is *image mosaicing*, which is the alignment of multiple images into a single composition representing part of a 3D scene. Various methods of presenting the composite image are demonstrated, and in particular, a novel algorithm is developed for automatically choosing an optimal viewing transformation for certain cases. In addition, a new and efficient method is demonstrated for the matching of point features across multiple views.

The second frame-fusion method is *super-resolution*, which aims to restore poor-quality video sequences by removing the degradations inherent in the imaging process. The framework presented here uses a generative model of the imaging process, which is discussed in detail and an efficient implementation presented. An algorithm is developed which seeks a maximum likelihood estimate under this model, and the factors affecting its performance are investigated analytically and empirically.

The use of "generic" prior image models in a Bayesian framework is described and shown to produce dramatically improved super-resolution results. Finally, super-resolution algorithms are developed which make use of image models which are tuned to specific classes of image. These algorithms are shown to produce results of comparable or better quality than those using generic priors, while having a lower computational complexity. The technique is applied to images of text and faces.

Throughout this work, the performance of the algorithms is evaluated using real image sequences. The applications demonstrated include the separation of latent marks from cluttered, non-periodic backgrounds in forensic images; the automatic creation of full 360° panoramic mosaics; and the super-resolution restoration of various scenes, including text and faces in low-quality video.

Acknowledgements

There are a great many friends and colleagues to whom I am indebted for the parts that they have played in the production of this book. I am extremely grateful to my supervisor, Professor Andrew Zisserman. I have benefitted tremendously from his considerable experience, enthusiasm, encouragement and advice, offered freely and generously throughout the years that we have worked together. I also extend my sincere thanks to Dr. Andrew Fitzgibbon for his guidance and assistance in all areas of my research. I would like to thank my examiners, Dr. Stephen Roberts and Dr. Michal Irani, for their many helpful comments and suggestions. I am also grateful to Simon Bramble and David Compton of the Forensic Science Service for supplying the latent mark images. I would especially like to thank Beth whose love, understanding and company were a great comfort throughout the dark days of thesis-writing. Finally, and perhaps most importantly, I am forever grateful to my parents for their encouragement and support in all my endeavours.

Contents

Chapter 1

Introduction

This book investigates sets of images consisting of many overlapping views of a scene, and how the information contained within them may be combined to produce single images of superior quality. The generic name for such techniques is *frame fusion.* Using frame fusion, it is possible to extend the field of view beyond that of any single image, to reduce noise, to restore high-frequency content, and even to increase spatial resolution and dynamic range. The aim in this book is to develop efficient, robust and automated frame fusion algorithms which may be applied to real image sequences.

An essential step required to enable frame fusion is *image registration*: computing the point-to-point mapping between images in their overlapping region. This subproblem is considered in detail, and a robust and efficient solution is proposed and its accuracy evaluated. Two forms of frame fusion are then considered: *image mosaicing* and *super-resolution.* Image mosaicing is the alignment of multiple images into a large composition which represents part of a 3D scene. Super-resolution is a more sophisticated technique which aims to restore poor-quality video sequences by modelling and removing the degradations inherent in the imaging process, such as noise, blur and spatial-sampling.

A key element in this book is the assumption of a *completely uncalibrated camera.* No prior knowledge of the camera parameters, its motion, optics or photometric characteristics is assumed. The power of the methods is illustrated with many real image sequence examples.

1.1 Background

The camera is a device which measures scene intensities and, just like any other measuring instrument, has a transfer function which introduces information loss into the measurement process. Bandwidth reduction, quantization, frequency aliasing and noise are common degradations found in imaging systems. Consequently, the images are often unable to completely capture the fine detail in a scene.

Image restoration techniques use a model of the imaging process which is "deconvolved" from the measured images in an attempt to recover the undegraded scene

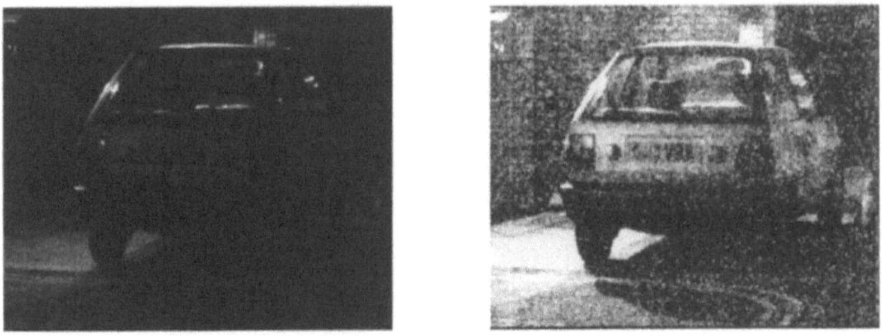

Figure 1.1: (Left) A single frame from a video sequence of a static scene taken under very poor lighting conditions. (Right) After histogram equalization. The license plate is still unreadable due to the noisy nature of the image.

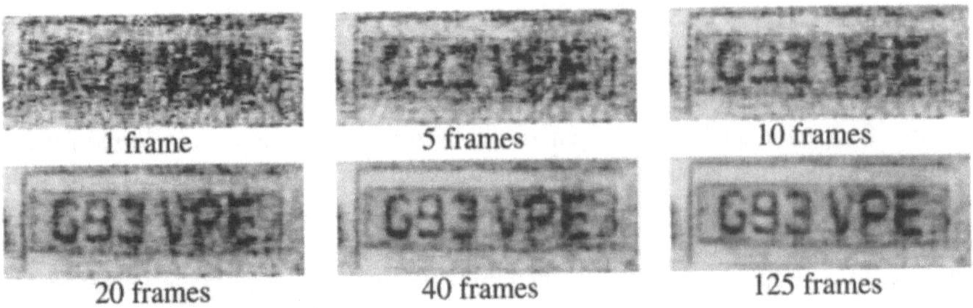

Figure 1.2: Close-up of the license plate of Figure 1.1 (histogram equalized). Each image is the result of averaging together an increasingly large number of frames. Eventually the text becomes readable.

intensities. Traditional methods have only applied to single images and are therefore fundamentally limited by the inherent loss of information in the imaging process. The restoration problem is ill-posed (under-constrained) and the techniques must rely heavily on prior assumptions about the scene intensities.

By combining information from multiple images of the same scene, the number of constraints on the reconstructed image can be greatly increased, thus improving the condition of the problem and reducing the extent to which a strong prior model of the image is required.

Possibly the simplest example of frame fusion is temporal averaging, which is an extremely effective means of reducing image noise in video sequences of static scenes. Figure 1.1 shows a frame from a video sequence of a stationary car captured under very poor lighting conditions with a fixed camera. Histogram equalization of a single frame fails to reveal the license plate due to the extremely noisy nature of the images. However, as an increasing number of frames are averaged together, the noise is reduced and the plate eventually becomes clearly readable, as shown in Figure 1.2.

The range of problems to which frame fusion is applicable is greatly increased when the single view-point constraint is removed. If multiple cameras/views are to be used, there is then the requirement that every image be accurately mapped into some global reference frame, or equivalently, that the same scene point may be accurately

Figure 1.3: (Top) 8 frames from a sequence of 25 captured by a hand-held video camera. (Bottom) All 25 frames are combined by image mosaicing. The field of view is greatly increased.

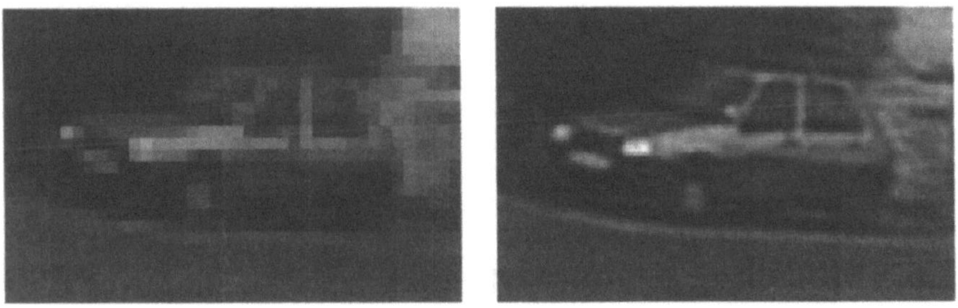

Figure 1.4: (Left) A region-of-interest in an image captured by a hand-held Mini-DV camera. (Right) A super-resolution estimate of the underlying scene combining information from 50 such images. The reconstruction is at 3× the original resolution.

located in every image in which it appears. The process of computing these mappings is called *image registration*. Having registered a set of images with a common frame of reference, there are two principal forms of frame fusion which may then be applied.

Image mosaicing The images themselves may be geometrically warped and combined in a manner which both reduces noise and greatly increases the effective field of view. This is known as *image mosaicing*, and it may be used to compose tens or hundreds of images into wide-angle, panoramic views, such as that shown in Figure 1.3, which is composed from 25 images.

Super-resolution If the image registration is of high enough accuracy, the many overlapping images may be used to increase the spatial sampling density of the scene, thus allowing recovery of image frequencies above the Nyquist limit of any single image. This *super-resolution* idea is one of the main topics of this book . Figure 1.4 shows a super-resolution reconstruction which combines information from 50 images into a single, still image at 3× the original resolution.

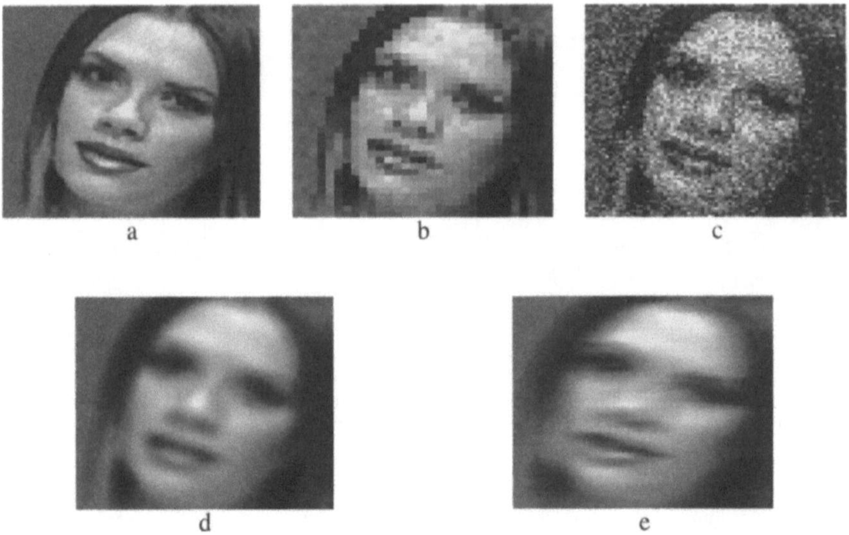

Figure 1.5: The effects of various imaging degradations. (a) An undegraded image. (b) The result of subsampling the image. (c) Noise is introduced to the intensity values. (d) The original image is subject to optical blur. (e) Motion in the scene causes anisotropic blurring.

1.2 Modelling assumptions

The information loss in camera images is due to a combination of different image degradations, each of which is characterized by a particular transfer function. The highest spatial frequency which can be captured by the camera is limited by the resolution of its imaging transducer - usually a *charge-coupled device* (CCD) array. This leads to spatial quantization of the image, which is illustrated in Figure 1.5(b). The CCD array is subject to various sources of noise, including thermal noise, shot noise, and electronic noise in the amplifier circuitry. This is particularly evident in low-lighting conditions when the camera gain is very high (see Figure 1.5(c)). When the image is digitized it also suffers intensity quantization, usually to 8 bits of precision.

It is common to consider the camera to be the familiar *pin-hole camera*, in which case everything is always in focus. In reality, of course, this is not the case and the image is therefore subject to optical blur which occurs prior to the spatial quantization. The effect of optical blur is illustrated in Figure 1.5(d). Finally, if the scene is moving, the energy entering the camera from a point in the scene will be integrated over several CCD cells during the shutter time of the camera. This causes motion blur, an example of which is shown in Figure 1.5(e).

For the purposes of image registration, this book is concerned only with images that are related by an 8 degree-of-freedom (dof) plane projective transformation (or *homography*), for which we can automatically compute the dense image-to-image correspondence. This motion model is suitable for images of a planar scene, and also for images taken by a stationary camera rotating about its optic centre.

1.3 Applications

Image mosaicing has already made an impact on the "prosumer" digital photography market, with the emergence of several products which allow a handful of photos or even a video stream from a hand-held camera to be stitched together into a wide-field mosaic, such as Peleg and Herman's VideoBrush system [112]. Mosaicing also forms the basic technology underlying node-based virtual reality systems such as QuickTime VR [1, 35] and SmoothMove [5]. The technique is also finding applications in video compression, digital steady-cam, and in the enhancement of digital video editing and matte placement software.

Super-resolution from uncalibrated video is still a relatively young field, and algorithms that are robust enough to be applied to real image data are only now beginning to emerge. The most obvious area of application is in forensic image processing. In the recent trial of the beating of Reginald Denny in the 1992 Los Angeles riots, one of the assailants was uniquely identified by a rose-shaped tattoo, enhanced in a single image using Cognitech's Video Investigator software [2]. The success of such cases had led to the rapidly increasing acceptance of digitally restored images in courtroom evidence. One fairly successful non-forensic application is the Salient Stills software [4] which attempts to generate photo-quality still images from low-quality, interlaced video streams. More recently emerging applications include resolution enhancement of mosaic images, and the generation of high-quality texture and environment maps for use in computer-generated special effects.

1.4 Principal contributions

The remaining chapters and their principal contributions are as follows.

Chapter 2: Literature survey

- A detailed survey is made of the literature pertaining to accurate image registration, image mosaicing and spatial-domain methods for uncalibrated super-resolution.

Chapter 3: Registration: geometric and photometric

- Robust methods are described for the accurate geometric and photometric registration of images, and their performance is demonstrated using real images. The accuracy of the geometric registration algorithm is investigated empirically.

- The registration methods are applied to forensic images in an application which allows latent marks to be separated from confusing, non-periodic backgrounds. The success of the method is demonstrated by several examples.

Chapter 4: Image mosaicing

- A novel algorithm is described for the efficient matching of image features across multiple views which are related by projective transformations.
- An automatic method is demonstrated for choosing an optimal viewing transformation, which aims to minimize projective distortion in the rendered mosaics.

Chapter 5: Super-resolution: Maximum likelihood and related approaches

- A detailed discussion is given of the implementation and efficiency issues regarding the generative imaging model used in super-resolution reconstruction. An effective and efficient implementation is described.
- The behaviour of the maximum-likelihood super-resolution estimator, its sensitivity to observation noise and modelling error, is explored both empirically and analytically. The method is compared in detail with Irani and Peleg's classic super-resolution algorithm [84].

Chapter 6: Super-resolution using Bayesian priors

- The use of generic image priors in a Bayesian approach to super-resolution is discussed. The performance of several such models is investigated empirically. A novel algorithm is described which uses cross-validation to determine the optimal weighting to be given to the prior in these estimators.

Chapter 7: Super-resolution using sub-space models

- A novel approach to super-resolution is described which uses sub-space image models, both in ML and MAP estimation. In some cases, the range-space of these models may be learnt from training data, and thus tuned to a particular class of image. The compactness of the model has a regularizing effect on the reconstruction problem, without imposing undesirable smoothness on the solution. The approach is applied to the enhancement of text and human faces, and is shown to be comparable or superior to methods using generic priors.

Chapter 8: Conclusions and extensions

- Several possible avenues of future research are discussed in detail.

An unfortunate feature of much of the research into uncalibrated super-resolution is the failure of some authors to present any results of their algorithms applied to real image sequences, opting instead to use purely synthetic data. In contrast, all of the algorithms presented in this book are demonstrated by application to real image sequences.

Chapter 2

Literature Survey

This chapter presents a summary of the literature relevant to the frame fusion techniques investigated in this book . It is broken down into three sections: image registration, image mosaicing and super-resolution.

2.1 Image registration

Essential to the work on mosaicing and super-resolution estimation is the need to find an accurate point-to-point correspondence between images in the input sequence. This is known as *registering* the images. The correspondence problem can be stated as follows: *given two different views of the same scene, for each image point in one view find the image point in the second view which has the same* pre-image, *i.e. corresponds to the same actual point in the scene.* For general scenes this mapping is complex and must be explicitly defined at every image point. However, under certain conditions the mapping can be expressed as a simple geometric relationship between the images.

Methods for obtaining registration fall into two broad categories : *direct methods* that compute a transformation which optimizes some measure of photometric consistency over the whole image; and *feature-based methods* that use a sparse set of corresponding image features (e.g. points and lines) to estimate the image-to-image mapping.

2.1.1 Registration by a geometric transformation

Of particular interest is the imaging of plane surfaces under general camera motion. In this case, the images are related by an 8 degree of freedom (dof) *planar projective transformation*, also called a *homography.* Methods for computing this and other transformations given a pair of images are now reviewed.

Which transformation? The most general case of the transformation is the 8 dof planar homography. For special camera motions the mapping between images of planes is simpler than a general homography. These simpler mappings have fewer dof, and consequently can often be more reliably and rapidly estimated.

If the camera motion consists of a translation parallel to the image plane and rotation about the principal axis then images are related by a 4 dof *similarity transformation*. This motion is often the case in satellite imaging and document scanning [34, 132]. Under tele-photo viewing conditions, where perspective effects such as the convergence of imaged parallel lines are negligible, images are related by a 6 dof *affine transformation* [34, 82].

Both similarity and affinity mappings are perfectly correct and valid transformations under certain imaging conditions, they are subgroups of a planar homography. By contrast, some authors have worked with more general imaging conditions in which only the full homography can correctly capture the image transformation, but for reasons of numerical stability and/or simplicity of optimization have chosen to approximate the homography by a Taylor expansion to second-order, resulting in the *biquadratic transformation* [93, 97, 98, 172]. This has 12 dof, but is unable to correctly model perspective effects. The expansion is given in the following equations.

$$\begin{aligned} x' &= q_{x'x^2}x^2 + q_{x'xy}xy + q_{x'y^2}y^2 + q_{x'x}x + q_{x'y}y + q_{x'} \\ y' &= q_{y'x^2}x^2 + q_{y'xy}xy + q_{y'y^2}y^2 + q_{y'x}x + q_{y'y}y + q_{y'} \end{aligned}$$

In some cases this expansion has allowed the authors to avoid expensive non-linear optimization algorithms which are generally necessary in computation of the exact homography, but claims of numerical instability in these methods appear to be unfounded (see chapter 3).

Which computation method? As explained in chapter 3, to compute a homography between two images only requires four or more point correspondences. For certain types of images these point correspondences can be obtained automatically by matching image features such as interest points (such as generated by a Harris corner detector) [13], and employing efficient non-linear optimization techniques to compute the required transformation. These feature-based methods have proved time and again to be extremely accurate, robust and efficient for the computation of many geometric image relations. Examples are [75, 154, 155, 157–159]. The challenge in implementing these algorithms is in the accurate and reliable detection of image features, and in the robust and efficient matching of corresponding features in two or more views. Feature-based methods are used throughout this book , and are discussed in more detail in chapter 3.

For reasons which are largely to do with available software and hardware in many labs, rather than performance, algorithms for computing homographies have generally been based on Gaussian or Laplacian pyramid, multi-scale approaches, often using simple gradient descent schemes to perform optimization.

The idea is that matching "large scale" features is easier (less searching is required). The transformation can then be followed through a scale space. Suppose the homography between two images I_1 and I_2 is sought. Typically a Gaussian pyramid is constructed for both I_1 and I_2 so that each level consists of progressively subsampled images. Matching begins at a coarse level of the pyramid: one of the images, I_1 say, is warped and the difference between the warped image and I_2 at that level measured. The transformation is found that minimizes this difference (a simple search, as there are not too many pixels, and the cost function should be smooth). This transformation is then applied to the next level of the pyramid, and refined at that level. The process repeats until the full image resolution is reached.

Irani *et al.* [82] use a multi-scale method to compute affine and quadratic transformations between images. Their "scale-space" is a Laplacian pyramid in which they minimize the sum of squared errors between registered images using a non-linear minimizer to estimate the parameters of the transformations. Mann and Picard [98] also use a multi-scale approach to estimate a quadratic approximation by minimizing an error based on projective flow. The multi-scale approach allows these algorithms to work for large image disparities.

Szeliski [145, 147] computes the full 8 dof homography between images using a multi-scale, iterative method. He uses a Gaussian pyramid and the Levenberg–Marquardt [116] algorithm to perform the estimation (see also Appendix A). He also suggests using matched image features to obtain an initial registration.

A modification of these methods which claims to produce very accurate registration is that of Bober *et al.* [18]. Their technique involves identifying areas of the images with rich texture and minimizing a correlation score between these areas in order to improve the estimate of the homography. By concentrating on "information rich" areas of the image their method represents an improvement in speed and noise sensitivity over other correlation-based algorithms.

2.1.2 Ensuring global consistency

The registration methods described so far are designed to compute the homography between pairs of overlapping images. In situations featuring many views of a scene, such as is the case in this book , it is often required to compute the homography relating any given pair of images (i, j). One way to achieve this would be to attempt registration between every possible pair of images in the sequence. But in practice, particularly for large numbers of images, this is impractical. A more common method is to compute homographies only between temporally consecutive images in a sequence, and then use the concatenation (multiplication) property of homographies to obtain the transformation between non-temporally consecutive views. However, this method is prone to "dead-reckoning" error which accumulates when concatenating transformations over long sequences.

Several authors have addressed this problem, attempting to compute a *globally consistent* set of homographies, such that the transformation between distant images (i, j) obtained by concatenation is equal to that which would be computed by direct registration of i and j.

Davis [45] suggests a method whereby a redundant set of homographies is first computed by registering all consecutive and some non-consecutive views. He then minimizes an algebraic residual based on the actual parameters of the homographies, attempting to ensure that the same homography $\mathtt{H}_{ij}$ computed by different concatenation paths is globally consistent. Although simple and easy to implement, the algebraic residual used does not correspond to any meaningful geometric error.

Hartley [75] extends the feature-based method for two-view homography computation to the case of *simultaneous* estimation of homographies over N-views, referring to [140] to explain how block matrix methods may be employed in order to render the required non-linear optimization tractable (known as *block bundle-adjustment*). This method is guaranteed to produce globally consistent homographies. It is shown to be efficient and accurate, but the problem of how to match corresponding feature points across many views is not addressed.

Attempts at global consistency have also been made using direct methods. Sawhney and Kumar [127] propose a scheme whereby, having obtained initial pairwise consecutive transformations, additional registration is performed between image pairs which are deemed to be "spatially" adjacent, i.e. they image the same part of the scene. Homographies between any pairs (i, j) are obtained by concatenating along the shortest path through the "view-graph" so formed. Although this method can improve consistency to an extent, the homographies are still computed using a pair-wise algorithm, hence global consistency is not achieved. They also propose a method whereby consecutive images are registered one at a time with a mosaic representation of the scene (see chapter 4), updating the mosaic with each new image. Again, this method does not produce globally consistent homographies.

More recently, Zelnik-Manor and Irani [172, 173] propose a scheme which does aim to simultaneously optimize N homographies, though their method is limited to small perturbations of the camera, since it uses the local biquadratic approximation to the full projective motion model.

2.1.3 Other parametric surfaces

The essence of this section is that images can be registered by a global mapping with a small number of parameters for planar scenes. There are other situations where the mapping is global but has more parameters. An example is an imaged quadric surface where the mapping has nine parameters and can be determined from nine correspondences [43]. This would allow simple mosaicing for images of curved surfaces such as spheres, cylinders, ellipsoids. A similar technique can be applied in general to surfaces which can be computed from image correspondences. This requires that the surface is known, or it lies in a specific class (for example, quadric, cubic, surface of revolution etc).

2.2 Image mosaicing

Image mosaicing is the alignment of multiple images into larger compositions which represent portions of a 3D scene. It generally applies to images related by planar homographies [35, 98, 147]. The conditions under which this mapping is applicable are described in detail in chapter 3, the most common being views of a plane from arbitrary camera positions, or views of a general scene taken by a camera free only to pan, tilt and zoom. Building a planar mosaic requires the images to be warped, using the computed homographies, into a common coordinate frame, and combined to form a single image. In this way it is possible to build panoramic views of scenes too large to fit into a single image, or alternatively to store the single mosaic image and use it to render views with arbitrary gaze directions and camera zoom.

Excellent reviews of image mosaicing and its many applications are given in [81, 82, 93]. Some notable examples are due to Kang and Szeliski [89, 90] who use mosaics composed of a hemisphere of images to represent the view in every direction at a particular point in the world. By capturing mosaics at many points, and matching image features across mosaics, they are used to perform wide-baseline 3D scene reconstruction. Coorg and Teller [37, 38, 148] use hemispherical mosaics in a similar application. The QuickTime VR software [35] is another example of the use of spherical mosaics to represent the view in every direction from some point in space. The representation is used to render the view in an arbitrary direction, with arbitrary zoom, thus creating a simple virtual reality application.

Other uses of mosaics include the creation of "clean-plates", high-quality images of a scene or texture from which independently moving objects have been removed. Such images are used for texture and environment mapping in the computer-generated special effects industry. A simple way to achieve this result is by temporal median filtering [81]. Davis [45] proposes a more sophisticated method of rendering mosaics that are free of artifacts caused by independently moving objects.

Some authors have attempted to extend mosaicing to more general scenes and camera motions with varying degrees of success. Some of these methods are broadly based on simulating the linear push-broom cameras used in satellite imaging. Rousso [120, 121] and Peleg [108, 111, 114] consider mosaics composed of strips extracted from the input images. The strips are chosen such that the direction of optic flow is orthogonal to the axis of the strip. Using this method, and with suitable blending, it is possible to make approximate mosaics for situations including camera translation, both along the optic axis and parallel to the image plane. Recently, Peleg has extended this technique to the creation of stereo-pairs from a single rotating camera. This modification requires that the camera be rotated about a point some distance from the optic centre. Mosaics composed using strips extracted from the left- and right-hand extrema of each image are used to form the stereo pair of mosaics [109, 110].

Recent interests in image mosaicing have included the creation of images with increased dynamic range, achieved by combining images taken under multiple different exposures. Two examples of this are due to Aggarwal and Ahuja [7] and Schechner and Nayar [129]. Both of these methods cleverly achieve the effect of multiple exposures by placing a spatially varying filter in front of the lens. Thus, as the camera rotates, each scene point is imaged with a varying level of attenuation.

Other methods for non-planar scenes Many other mosaicing related methods have been proposed for the efficient representation of scenes which deviate marginally from the zero-parallax model. One such method is Adelson's "layers" technique [6] which involves segmenting the image sequence into multiple independent planar motions, each described by a planar mosaic and transparency map. Adelson uses binary transparency maps, whilst Irani and Peleg [85] extended the approach to real-valued transparencies to cope with transparent motions. The "planes plus parallax" technique [93] attempts to overcome the limitations of planar mosaicing by describing the image motion between successive frames as a homography and a parallax field which is defined for every point in the image and accounts for the residual motion of points lying off the plane. This is effectively a general dense correspondence problem. If the camera centres are co-linear, it is possible to align the parallax fields and reproject the images into a common reference frame as if they were taken *by a camera rotating about its optic centre*. Co-linear cameras share the same pencil of epipolar planes and can hence be made to share the same epipolar lines by means of an appropriate homography (a synthetic rotation about the camera centre.) The parallax field in this case simply encodes the disparity of points along their epipolar lines. A very similar technique has been documented in the computer graphics community under the name of "view interpolation" (see [36, 134]).

2.3 Super-resolution

Historically, there have been a variety of approaches to the super-resolution restoration of images. Early attempts were based on the generalized sampling theorem [167]. The drawback with these methods is that they do not allow for blur or noise in the observed images. Another popular approach is to perform restoration in the frequency domain [152, 163]. These methods can deal with spatially invariant blur and noisy observations, but the motion model is limited to pure translation. More recent approaches have used *projection onto convex sets* (POCS) to solve iteratively the super-resolution inverse problem using a full generative image model and arbitrary motion model [55, 58, 59, 105–107]. POCS is a very simple way to impose constraints such as non-negativity and bounded errors. However, POCS-based approaches to super-resolution suffer from extremely slow convergence, and because they optimize a purely constraint-based objective, they do not converge to a unique solution.

The methods investigated in this book tackle the super-resolution inverse problem in the spatial domain using a statistical framework. Bayesian methods are employed to regularize the problem, and constraints are imposed in a computationally efficient manner. This allows maximum flexibility in the generative image model and motion model whilst allowing very efficient optimization strategies. The unique solutions obtained are optimal in a plausible statistical sense.

2.3.1 Simple super-resolution schemes

Enhancement by frame-fusion is possible when we have several images of the same surface and a dense correspondence (from either geometric or general registration) between the images. It attempts to reverse all of the degradations at once, recovering a deblurred, high spatial-resolution surface texture.

The simplest methods attempt to resample all the observed image data into a single coordinate frame. The resampled data is merged by averaging or median filtering to obtain a low-noise, high-density image. A standard, single-image deblurring step completes the process.

Early work was done by Ur and Gross [167], based on generalized sampling theory. They assume that the input images undergo relative shifts which are known precisely *a priori*. The low-resolution images are interpolated and merged onto a finer grid, before deblurring by convolution with a kernel derived from the inverse of the blur operator. A similar technique has been proposed by Rudin *et al.* [122]. They use a hierarchical block-matching algorithm to obtain a dense optic flow field. One view is chosen as a reference frame and its resolution is increased to the desired level using a simple interpolation kernel. The other images are warped and merged into the reference frame according to the optic flow field. Finally, a standard single-image deblurring algorithm is applied to obtain the super-resolution result. We examine this algorithm in more detail in Chapter 5.

2.3.2 Methods using a generative model

More recent methods use a *generative model* of the camera transfer function which determines how a real surface is transformed, filtered and sampled to form an image; and also an accurate set of registrations between the input images. They proceed by finding a high-resolution image which, when transformed according to the registration parameters and degraded according to the camera transfer function, produces a set of simulated images which are as similar as possible to the actual observed low-resolution images.

Under the assumption of Gaussian distributed image noise, these methods produce the *maximum likelihood estimate* (MLE) of the super-resolved surface intensities. It is also common to impose a prior model of the surface intensities (such as a spatial-smoothness constraint), in which case the intensity estimates obtained are the *maximum a posterior estimates* (MAP).

Algorithms proposed vary in their registration methods, their use of prior texture models, and the numerical methods used to converge to the required estimate. Irani and Peleg [84, 85] consider images obtained from a flatbed scanner which have undergone rotation and translation (a Euclidean transformation). Their registration method is based on a coarse-to-fine, texture correlation strategy, as described in chapter 3. They consider optical blur, and obtain the kernel of the point spread function (PSF) by imaging a small dot. Their cost function to be minimized is the sum of squared differences in intensity values between the simulated low-resolution images and the actual ones (see Equation (2.1) in which $\hat{g}_n$ is the nth simulated image and g_n the associated actual low-resolution image), and the super-resolved texture (typically double the resolution of the input images) is found by a simple iterative update scheme similar to steepest descent. The properties of this algorithm are examined in detail in chapter 5. Mann and Picard [97, 98] extended Irani and Peleg's algorithm to fully projective image registration, obtained using a coarse-to-fine texture correlation strategy.

$$\text{cost} = \sum_{n} \sum_{(x,y)} (\hat{g}_n(x,y) - g_n(x,y))^2 \tag{2.1}$$

Elad and Feuer [52–54, 56] also make use of a generative model within a POCS-based approach to super-resolution. They assume that accurate registration is known *a priori*, and fail to present any results based on real images. Tom and Katsaggelos [151] propose a frequency domain method based on a generative model, in which the registration (limited to translation) and super-resolution image are estimated simultaneously. Again, no real image results are presented.

Dellaert *et al.* [47] propose an on-line method for the continuous super-resolution update of a patch of texture as it is tracked. The affine tracking algorithm uses the current estimate of the patch as a template, and the super-resolution update is performed using a very simplified Kalman-style update scheme.

Shekarforoush and Challappa [136, 137] also propose several sequential update schemes for super-resolution estimation using a generative model. Additionally, they propose a novel method for estimating the blur function based on analysis of the cross-power spectrum of the images. Their method is applied to real image sequences, with unconvincing results.

2.3.3 Super-resolution using statistical prior image models

Bascle *et al.* [11] extended the back-projection approach to motion blurred images. The registration method involves tracking a region using a combined texture correlation and contour based approach related to snake tracking [12], assuming an affine motion model. Their cost function includes a second order smoothness constraint on the resolved texture, making this a MAP estimation method (see Equation (2.2) in which f is the high-resolution estimate). This ensures that the estimate is not dominated by periodic noise as is common when reconstructing motion blurred images. The texture estimate which minimizes the cost function is found by *conjugate gradient descent* [69].

$$\text{cost} = \sum_{n=0}^{N} \sum_{(x,y)} (\hat{g}_n(x,y) - g_n(x,y))^2 + \lambda \sum_{(x,y)} (\nabla^2 f(x,y)) \qquad (2.2)$$

Cheeseman *et al.* [34] consider satellite imagery in which the input images are assumed to be related by affine transformations. Registration again proceeds by a texture-based method, finding the transformation which minimizes the squared intensity error between registered images by gradient descent. They take a MAP approach, imposing a first order smoothness constraint in the reconstructed high-resolution image. In this case the PSF can be obtained from the bench calibration of the satellite optics.

Schultz and Stevenson [131] propose a MAP estimation method which differs from previous ones in two significant respects. Firstly, it does not use a geometric registration method, but a general full, dense correspondence method. Each pixel has a disparity vector associated with it which encodes its motion between the two images, and hence the method implicitly assumes that the scene is locally fronto-parallel at the pixel level. Secondly, the prior image model is based on a Huber-Markov random field (HMRF) which represents piecewise-smooth data. The smoothness constraint imposed by the methods discussed in previous paragraphs is based on a Gauss-Markov image model which imposes an x^2 penalty on the image gradient. The HMRF model behaves this way for small gradients, but imposes a less severe penalty at large gradient discontinuities. It is therefore much better at preserving edges in the super-resolved image. Again, a suitable estimate is found by a gradient descent method.

Very recent work has focused on inferring super-resolution images from a single observed low-resolution image. Single-image methods are inherently limited by the amount of data available in an image and by the noise present in the image. However, Candocia and Principe [26], Freeman and Pasztor [61–64], and Baker and Kanade [10] have proposed Bayesian MAP estimators which utilize sophisticated prior models learnt from training images. These methods attempt to infer high-frequency detail from a single low-resolution image.

Chapter 3

Registration: Geometric and Photometric

3.1 Introduction

This chapter describes the geometric and photometric registration techniques which enable the mosaicing and super-resolution methods presented in later chapters.

In the context of this book , *geometric registration* refers to the process of obtaining a dense correspondence (or registration) between multiple views of a planar surface, or equivalently, between multiple views taken by a camera rotating about its optic centre. In both cases, the geometric transformation between any two such views is captured completely by an 8 degree-of-freedom (dof) *planar projective transformation* or *homography*. Section 3.2 introduces the relevant geometry. In section 3.4, a robust estimation procedure for computing the registration parameters from corresponding points in multiple views is described, and in section 3.5 the accuracy of this method is investigated empirically.

Photometric registration refers to the procedure by which global photometric transformations between images are estimated and compensated for. Examples of such transformations are global illumination changes across the scene, and intensity variations due to camera automatic gain control or automatic white balancing. Section 3.7 presents a simple parametric model of these effects, along with a robust method for computing the parameters given a set of geometrically registered views. The method is shown to be effective when applied to real image sequences.

Finally, section 3.8, describes a real-life forensic science application which uses these registration algorithms to allow latent marks, such as fingerprints, to be separated from *non-periodic* background clutter, such as a bank-note. The application was developed as part of an industrial collaboration project and is now in use in the UK.

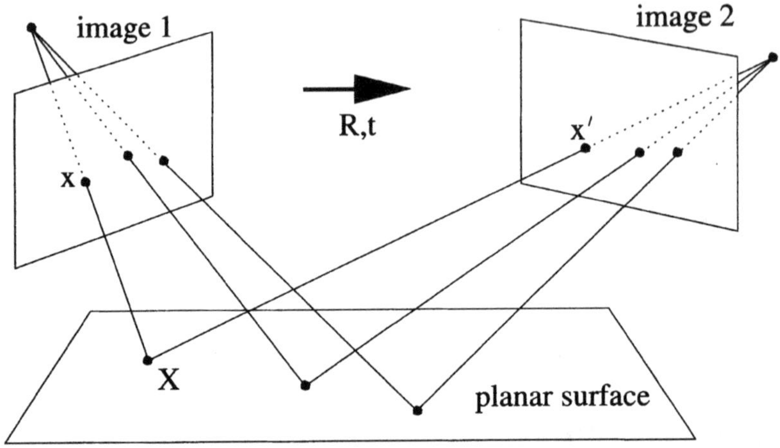

Figure 3.1: **Images of planes.** A planar homography is induced between the two images of a plane taken from different viewpoints (related by a rotation $\mathbf{R}$ and translation $\mathbf{t}$). The scene point $\mathbf{X}$ is projected to points $\mathbf{x}$ and $\mathbf{x}'$ in the two images. The image points are related by $\mathbf{x}' = \mathtt{H}\mathbf{x}$.

3.2 Imaging geometry

In this section, we briefly review the geometry of the planar homography, which is the mapping that arises in the perspective imaging of planes. This is the simplest case in which we can call on geometry to allow us to compute an accurate point-to-point mapping between two images. Geometric registration is also possible for multiple views of other parametric surfaces such as quadrics, where the mapping has nine parameters, but treatment of these surfaces and of general non-parametric surfaces is beyond the scope of this book . A detailed treatment is presented by Cross *et al.* [41–44], Shashua and Toelg [135], and Can *et al.* [25].

There are two important situations where the image-to-image mapping is exactly captured by a planar homography: images of a plane viewed under arbitrary camera motion; and images of an arbitrary 3D scene viewed by a camera rotating about its optic centre and/or zooming. The two situations are illustrated in Figures 3.1 and 3.2 respectively.

A third imaging situation in which a homography may be appropriate is due to a freely moving camera viewing a very distant scene, such as is the case in high-aerial or satellite photography. Because the distance of the scene from the camera is very much greater than the motion of the camera between views, the parallax effects caused by the three-dimensional nature of the scene are negligibly small.

In all cases it is assumed that images are obtained by a perspective pin-hole camera, one in which all rays between objects and their images intersect at the camera centre.

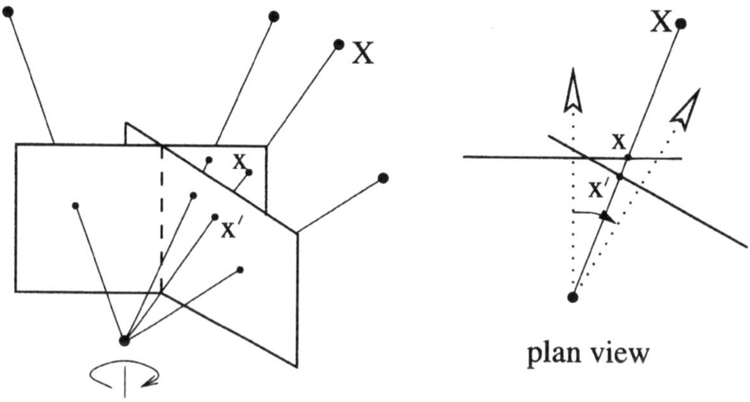

Figure 3.2: **Rotation about the camera centre.** As the camera is rotated the points of intersection of the rays with the image plane are related by a planar homography. Image points $\mathbf{x}$ and $\mathbf{x}'$ correspond to the same scene point $\mathbf{X}$. The image points are related by $\mathbf{x}' = \mathtt{H}\mathbf{x}$.

Notation Points are represented by homogeneous coordinates, so that a point (x, y) is represented as $(x, y, 1)$. Conversely, the point (x_1, x_2, x_3) in homogeneous coordinates corresponds to the inhomogeneous point $(x_1/x_3, x_2/x_3)$.

Definition: planar homography Under a planar homography (also called a plane projective transformation, collineation or projectivity) points are mapped as:

$$\begin{pmatrix} x'_1 \\ x'_2 \\ x'_3 \end{pmatrix} = \begin{bmatrix} h_{11} & h_{12} & h_{13} \\ h_{21} & h_{22} & h_{23} \\ h_{31} & h_{32} & h_{33} \end{bmatrix} \begin{pmatrix} x_1 \\ x_2 \\ x_3 \end{pmatrix} \tag{3.1}$$

or

$$\mathbf{x}' = \mathtt{H}\mathbf{x} \tag{3.2}$$

where the $=$ indicates equality up to a scale factor, and consequently the transformation matrix has 8 degrees of freedom.[1]

It is important to note that real cameras may deviate significantly from this pin-hole model. This gives rise to *radial distortion* at the periphery of the image, meaning that the mapping between views is not correctly modelled by a homography. The effect is chiefly observed when using cheap, wide-angle lenses. The sequences used in this book were captured using fairly high-quality equipment and are well modelled by the pin-hole camera. In any case, it is possible to correct this distortion, either *a priori* by imposing "straightness" of hand-selected line segments, or by including a simple parametric model of the lens distortion in the geometric registration procedure. The

[1] The matrix has eight degrees of freedom because only the ratio of homogeneous coordinates is significant. There are eight ratios among the nine elements of $\mathtt{H}$. Multiplying the matrix by an arbitrary non-zero number has no effect on this ratio.

former method is used in the context of single-view metrology by Criminisi *et al.* [39, 40]. The latter method is used in the context of image mosaicing by Sawhney and Kumar [127] and by Dellaert *et al.* [48]. A detailed analysis of the lens distortion problem is performed by Devernay and Faugeras [50], and also by Stein [143].

3.3 Estimating homographies

Various methods for computing a planar homography between image pairs have been proposed, but they generally fall into two broad categories :

- **Direct correlation methods** compute the homography by maximizing photometric consistency over the whole image.
- **Feature-based methods** compute the homography from a sparsely distributed set of point-to-point correspondences.

Almost exclusively, the results presented in this book were generated using feature-based registration methods. Feature-based techniques have many significant advantages over their direct correlation counterparts in terms of computation speed, and the scope that they offer for the application of robust statistical methods for outlier rejection. These differences are discussed in more detail in Section 3.6.

In this section we review the feature-based approach, and describe both linear and non-linear methods for estimating the homography relating two images.

3.3.1 Linear estimators

The planar homography has eight degrees of freedom. Each point correspondence generates two linear equations for the elements of $\mathtt{H}$ and hence four correspondences is enough to solve for the homography directly. If more than four points are available, a least-squares solution can be found by linear methods. We now briefly review the linear method for estimating the homography $\mathtt{H}$, from a set of N point correspondences $\mathbf{x} \leftrightarrow \mathbf{x}'$. Full details are given in Hartley and Zisserman [76].

From the definition of $\mathtt{H}$, we have

$$\begin{pmatrix} x' \\ y' \\ 1 \end{pmatrix} = \begin{bmatrix} h_{11} & h_{12} & h_{13} \\ h_{21} & h_{22} & h_{23} \\ h_{31} & h_{32} & h_{33} \end{bmatrix} \begin{pmatrix} x \\ y \\ 1 \end{pmatrix}$$

where $=$ is equality up to scale. Each inhomogeneous, 2D point correspondence generates two linear equations in the elements of $\mathtt{H}$

$$\begin{aligned} x'\left(h_{31}x + h_{32}y + h_{33}\right) - h_{11}x - h_{12}y - h_{13} &= 0 \\ y'\left(h_{31}x + h_{32}y + h_{33}\right) - h_{21}x - h_{22}y - h_{23} &= 0 \end{aligned}$$

Hence, n points generate $2n$ linear equations, which may be arranged in a "design matrix" :

$$\mathtt{A}\mathbf{h} = 0 \tag{3.3}$$

The solution for $\mathbf{h}$ is the one-dimensional kernel of $\mathtt{A}$, which may obtained from the singular value decomposition (SVD). For $n > 4$ points, this equation will not have an exact solution. In this case, a solution may be obtained which minimizes the *algebraic residuals*, $\mathbf{r} = \mathtt{A}\mathbf{h}$, in a least-squares sense, by taking the singular vector corresponding to the smallest singular value.

3.3.2 Non-linear refinement

We have to assume that the measured positions of the image features used for the computation will be subject to noise. In this case it is important to ensure that using many correspondences will improve the estimate of the homography. The linear method provides a closed-form solution which is easy to compute. However, the algebraic residuals that are minimized do not correspond to a sensible geometric distance.

In order to obtain a unique estimate of $\mathtt{H}$ from noisy data we use the method of *maximum likelihood*. This provides an estimate of $\mathtt{H}$ which is optimal according to a plausible statistical model of the noise corrupting our observations. In this case, our observations are point-to-point correspondences, and each point $\mathbf{x}$ has a localization error which is distributed about the true point position $\underline{\mathbf{x}}$. Hence, intuitively, we would expect that a better estimate of $\mathtt{H}$ will be obtained if we minimize residuals which correspond to geometric distances in the images.

A common cost function is the sum of squared forward and reverse transfer distances,

$$C = \sum_i (d^2(\mathbf{x}'_i, \mathtt{H}\mathbf{x}_i) + d^2(\mathbf{x}_i, \mathtt{H}^{-1}\mathbf{x}'_i)) \tag{3.4}$$

where $(\mathbf{x}_i \leftrightarrow \mathbf{x}'_i)$ are the image correspondences, and $d^2(\mathbf{x}, \mathbf{x}')$ is the squared (inhomogeneous) Euclidean distance between homogeneous points $\mathbf{x}$ and $\mathbf{x}'$. The error metric is illustrated in Figure 3.3. This error can be minimized with respect to the 8 parameters of $\mathtt{H}$ using an iterative, non-linear least-squares optimizer. Efficient algorithms for performing this optimization are based on the Gauss–Newton method [49, 67, 68], the preferred variant being the Levenberg–Marquardt algorithm [116]. At every iteration, a finite difference approximation may be used to compute the Jacobian of the residuals with respect to each of the 8 parameters. This error metric has the advantage of being simple to compute and provides good estimates of the homography. However, it does not provide the *maximum likelihood estimate* of $\mathtt{H}$.

3.3.3 The maximum likelihood estimator of $\mathtt{H}$

To derive the error metric which must be minimized to give the MLE homography we model the localization error on the feature points as an isotropic, normal distribution with zero mean and standard deviation σ. These assumptions are verified in Section 3.5. Given a true, noise-free point $\underline{\mathbf{x}}$, the probability distribution of the corresponding feature point location is

$$\Pr(\mathbf{x}_i|\underline{\mathbf{x}}) = \frac{1}{2\pi\sigma^2} \exp\left(-\frac{(x-\underline{x})^2 + (y-\underline{y})^2}{2\sigma^2}\right) \tag{3.5}$$

Hence, given the true, noise-free correspondences $\{\underline{\mathbf{x}} \leftrightarrow \underline{\mathbf{x}}'\}$, and making the very reasonable assumption that the feature localization error is uncorrelated across different images, the probability density of the observed, noisy correspondences $\{\mathbf{x} \leftrightarrow \mathbf{x}'\}$ is

$$\Pr(\{\mathbf{x}, \mathbf{x}'\}) = \prod_i \Pr(\mathbf{x}_i|\underline{\mathbf{x}}_i) \Pr(\mathbf{x}'_i|\underline{\mathbf{x}}'_i) \tag{3.6}$$

The negative log-likelihood of the set of all correspondences is therefore

$$\begin{aligned} L &= -\sum_i (\log \Pr(\mathbf{x}_i|\underline{\mathbf{x}}_i) + \log \Pr(\mathbf{x}'_i|\underline{\mathbf{x}}'_i)) & (3.7) \\ &= \sum_i ((x_i - \underline{x}_i)^2 + (y_i - \underline{y}_i)^2 + (x'_i - \underline{x}'_i)^2 + (y'_i - \underline{y}'_i)^2) & (3.8) \end{aligned}$$

Of course, the true pre-image points are unknown, so we replace $\{\underline{\mathbf{x}}, \underline{\mathbf{x}}'\}$ in the above equation with $\{\hat{\mathbf{x}}, \hat{\mathbf{x}}'\}$, the estimated positions of the pre-image points, hence

$$L = -\sum_i ((x_i - \hat{x}_i)^2 + (y_i - \hat{y}_i)^2 + (x'_i - \hat{x}'_i)^2 + (y'_i - \hat{y}'_i)^2) \tag{3.9}$$

Finally, we impose the constraint that $\hat{\mathbf{x}}$ maps to $\hat{\mathbf{x}}'$ under a homography, and hence substitute

$$\hat{x}' = \frac{h_{11}\hat{x} + h_{12}\hat{y} + h_{13}}{h_{31}\hat{x} + h_{32}\hat{y} + h_{33}}, \quad \hat{y}' = \frac{h_{21}\hat{x} + h_{22}\hat{y} + h_{23}}{h_{31}\hat{x} + h_{32}\hat{y} + h_{33}}$$

This error metric is illustrated in Figure 3.3. The required homography and pre-image point estimates $\{\hat{\mathbf{x}}\}$ minimize this function. A direct method of obtaining these estimates is to parameterize *both* the eight parameters of the homography, *and* the $2N$ parameters of the N points $\{\hat{\mathbf{x}}\}$. However, this optimization requires special consideration if it is to be done efficiently. We return to this idea in the context of homography computation over multiple views in Section 4.4. Fortunately, in the two-view case, it is possible to derive a very good approximation to this log-likelihood function, based on Sampson's approximation of the distance to a conic [123], which does not require explicit parameterization of the pre-image points. The details of this approximation are described by Torr *et al.* [156, 160]. Again, the Levenberg–Marquardt algorithm is used to optimize the approximate log-likelihood with respect to the 8 homography parameters, yielding a very accurate estimate of $\mathtt{H}_{\mathrm{mle}}$.

3.4 A practical two-view method

The automatic, feature-based method used throughout this work to compute accurate homographies is summarized in Table 3.1. As a practical example, we consider the image pair shown in Plate 1.

Feature extraction The point features used are extracted using the Harris feature detector [73] which automatically extracts hundreds of stable image features (known as interest points or "corners") with positions accurate to around 0.25 pixels. However,

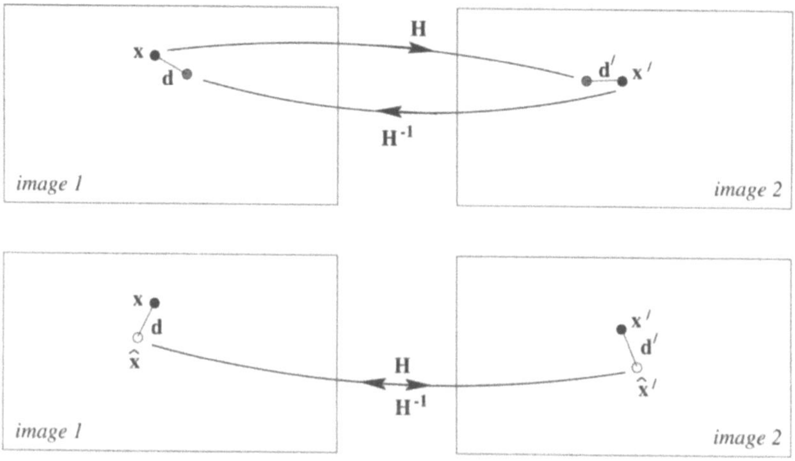

Figure 3.3: (Top) Homography computation using the forward-backward transfer distance metric of Equation (3.4). The metric is simple to use, but does not give the maximum likelihood (ML) estimate of H. (Bottom) The ML estimator of Equation (3.9) minimizes the reprojection error between the pre-image point correspondence $(\hat{\mathbf{x}}, \hat{\mathbf{x}}')$ and the observed interest points $(\mathbf{x}, \mathbf{x}')$.

as can be seen from Plate 1, the term "corner" is misleading, as these point features do not only occur at real corners (line intersections). Thus the term *interest point* is preferred. There will typically be hundreds or thousands of interest points detected in an image, depending on the image complexity.

Putative correspondences For every interest point $\mathbf{x}$ in the first image (Plate 1c), a match is sought amongst a subset $\mathcal{S}$ of the interest points in the second image. The subset $\mathcal{S}$ is defined by considering only the interest points $\{\mathbf{x}'\}$ contained within a search window centred on the location of $\mathbf{x}$ (Plate 1d). Every interest point $\mathbf{x}' \in \mathcal{S}$ is ranked by computing the normalized cross-correlation score between 5×5 pixel neighbourhoods centred on $\mathbf{x}$ and $\mathbf{x}'$ (Plate 1 e,f). Pairs whose correlation score falls below a conservative threshold are disregarded, and among the remaining pairs (if any) a winner-takes-all strategy is employed whereby the top-ranking point is taken to be the putative match for $\mathbf{x}$. Plate 2a shows the resulting 268 matches superimposed upon the first image. There are clearly many mismatches at this stage.

RANSAC The RANdom SAmple Concensus algorithm, proposed by Fischler and Bolles [60], is an extremely simple and powerful method for estimating model parameters given a data set heavily contaminated with outliers to the correct model. It has proved extremely effective in sifting two-view relations, such as homographies and fundamental matrices, from point-to-point image correspondences, as amply demonstrated by Torr [153]. The method considers many random subsets of the data, each containing the minimum number of samples required to compute the model parameters exactly, and selecting the parameter set which has the largest number of inliers,

Objective Compute the 2D homography between two images.

Algorithm

1. **Features:** Compute interest point features in each image to sub-pixel accuracy (e.g. Harris corners [73]).
2. **Putative correspondences:** Compute a set of interest point matches based on proximity and similarity of their intensity neighbourhood.
3. **RANSAC robust estimation:** Repeat for N samples.
   (a) Select a random sample of four correspondences and compute the homography $\mathtt{H}$.
   (b) Calculate a geometric image distance error for each putative correspondence.
   (c) Compute the number of inliers consistent with $\mathtt{H}$ by the number of correspondences for which the distance error is less than a threshold.

   Choose the $\mathtt{H}$ with the largest number of inliers.
4. **Optimal estimation:** Re-estimate $\mathtt{H}$ from all correspondences classified as inliers, by maximizing the likelihood function (3.9) using a suitable numerical optimizer, such as the Levenberg–Marquardt algorithm [116].
5. **Guided matching:** Further interest point correspondences are now determined using the estimated $\mathtt{H}$ to define a search region about the transferred point position.

The last two steps can be iterated until the number of correspondences is stable.

Table 3.1: The main steps in the algorithm to automatically estimate a homography between two images using RANSAC and features. Further details are given in [76].

i.e. data points which are correctly mapped to within some threshold. In the case of fitting a homography, minimal sets of four point correspondences are extracted, $\mathtt{H}$ is computed using a linear method, and the inliers are defined as those matches $(\mathbf{x}, \mathbf{x}')$ for which $d(\mathtt{H}, \mathbf{x}, \mathbf{x}')$ - the Sampson distance between homogeneous points $\mathbf{x}$ and $\mathbf{x}'$ under $\mathtt{H}$ - is less than a distance threshold T (typically 1.25 pixels). This is repeated many times depending on the expected ratio of inliers to outliers. The final homography estimate is that with the highest inlier count. Plate 2b and 2c show the inliers and outliers resulting from RANSACing our 268 putative correspondences.

Optimal estimation The homography estimate returned by the RANSAC algorithm is used to initialize the non-linear maximum likelihood estimator described in Section 3.3.3. The set of point correspondences deemed to be inliers is used to compute a refined estimate of $\mathtt{H}$.

Guided matching The refined homography estimate returned by the optimal estimation stage is used to guide a search for more interest point correspondences. The search procedure is similar to that used to obtain the putative correspondences: given an interest point $\mathbf{x}$ in the first image, a match is sought in a search window centred on the expected position $\mathbf{x}' = \mathtt{H}\mathbf{x}$ in the second image. Because the search is now guided, the size of the search window can be greatly reduced, typically 3×3 pixels, and the correlation threshold below which matches are discarded may be increased. The new set of inliers are again used to refine the optimal estimate of $\mathtt{H}$. The optimal estimation and guided matching stages are repeated until the number of inliers stabilizes. The final set of inlying correspondences is shown in Plate 2d.

There are two important things to note about the algorithm. First, interest points are not matched purely using geometry – i.e. only using a point's position. Instead, the intensity neighbourhood of the interest point is also used to rank possible matches via the normalized cross-correlation between the point's neighbourhood and the neighbourhood of each possible match. Second, the use of robust statistics is essential to the success of the algorithm: in the example above, more than 40% of the putative matches between the interest points (obtained by the best cross-correlation score and proximity) are incorrect. It is the RANSAC algorithm that identifies the correct correspondences.

3.5 Assessing the accuracy of registration

Super-resolution requires point-to-point mappings which are accurate to the sub-pixel level. In this section, we perform an empirical investigation to assess the accuracy of the feature-based method described in Section 3.4. A similar, though more detailed investigation was carried out by Schmid *et al.* [130], but with the emphasis on evaluating the performance of various feature detectors.

Clearly, the accuracy of feature location and registration can depend heavily on the image content and on the severity of the transformation. For example, images with strongly oriented textures are likely to yield feature points which are poorly localized in a particular direction, thus violating our assumption of isotropic error distribution. Furthermore, Harris corner localization is not invariant under scale change, again destroying isotropy. The purpose of this investigation therefore is to get an idea of the accuracy of our method applied to typical images undergoing moderately severe transformations.

3.5.1 Assessment criteria

The correct way to assess the quality of an estimated homography $\hat{\mathtt{H}}$ is to compare it with the ground-truth transformation $\underline{\mathtt{H}}$. Given such a ground-truth, we require a method of comparison which provides a meaningful and intuitive measure of the accuracy of the estimated homography. Simple, direct comparisons of the estimated parameters are of little value in this respect. What really interests us is "How accurately does this homography transform points between the two images?" In other words, we should base our quality criterion on the *transfer error* between the same point mapped under both $\underline{\mathtt{H}}$ and $\hat{\mathtt{H}}$:

$$\delta \mathbf{x}_1 = \underline{\mathtt{H}}(\mathbf{x}_1) - \hat{\mathtt{H}}(\mathbf{x}_1) \tag{3.10a}$$

$$\delta \mathbf{x}_2 = \underline{\mathtt{H}}^{-1}(\mathbf{x}_2) - \hat{\mathtt{H}}^{-1}(\mathbf{x}_2) \tag{3.10b}$$

where the notation $\mathtt{H}(\mathbf{x})$ indicates the mapping of an **inhomogeneous** 2D point $\mathbf{x}$ under a homography. The idea is illustrated in Figure 3.4. Note that we define the transfer error in both the forward and backward directions to maintain symmetry.

Given a ground-truth homography $\underline{\mathtt{H}}$ mapping from image 1 to image 2, and an estimate $\hat{\mathtt{H}}$, Equations (3.10a) and (3.10b) are used to compute a field of displacement vectors corresponding to the forward and backward transfer errors in each image. These fields are averaged in order to obtain the RMS transfer error of the estimated homography over a region of interest Ω :

$$e_{\mathrm{rms}} = \sqrt{\frac{1}{2\Omega}\int_{\Omega} (\delta \mathbf{x}_1)^2 + (\delta \mathbf{x}_2)^2 \, \mathrm{d}\Omega} \tag{3.11}$$

3.5.2 Obtaining a ground-truth homography

In practice, obtaining a ground-truth homography between a pair of real images is rather challenging. The method adopted here is to use a 17 inch LCD flat-panel screen

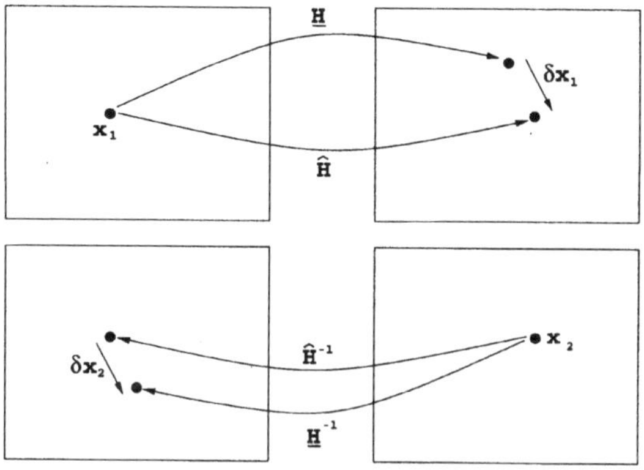

Figure 3.4: The forward and backward transfer errors are defined by mapping the same point $\mathbf{x}$ under the estimated and ground-truth homographies, $\hat{\mathtt{H}}$ and $\underline{\mathtt{H}}$.

to allow seamless switching between a calibration image and a set of test images. The calibration image is designed to allow very high accuracy homographies to be computed when it is viewed from different positions. For this purpose we use an 8×8 Tsai grid [164] with a $3:1$ mark–space ratio. Three different test images are then used: two Escher drawings ("Relativity" and "Mosaic") and a Lego scene. The calibration image and the "Lego" test image are shown in Figure 3.5.

The panel is viewed from four different viewpoints, and for each viewpoint, four images are captured: the calibration image and three different test-images. A mono Coho CCD camera is used for the image capture. The LCD panel has the advantage of being both extremely flat, and of allowing the image to be switched remotely, avoiding any risk of moving the experimental rig. Figure 3.6 shows the captured images.

High-accuracy homographies are computed between the four views of the calibration image by first fitting a template consisting of 16 horizontal and 16 vertical lines to each image. The algorithm for doing this is shown in Table 3.2. Having obtained accurate homographies between the template and each view of the calibration image, it is then a simple matter to obtain the homography between any pair of views.

Each line in the template is fitted to around 300 Canny "edgels", those which are deemed to be its inliers. The total number of residuals in this fitting process is therefore around $32 \times 300 = 9600$, depending of course on the particular view. The fitting error, which is indicated by the RMS orthogonal distance of the edgels to the fitted lines, is 0.02 pixels. It is reasonable to assume that the templates are fitted with sufficient accuracy to allow near ground-truth homographies to be computed. The fitted templates are shown in Figure 3.7.

Registration between the test images is performed using the feature-based, 2-view maximum likelihood estimator described in Section 3.4. Around 500 Harris interest points are extracted in each image, from which somewhere between 200 and 450 point correspondences are obtained, the exact number depending upon the particular test images in question. Figure 3.8 shows an example.

At this point, the reader may be wondering why we have gone to all this trouble

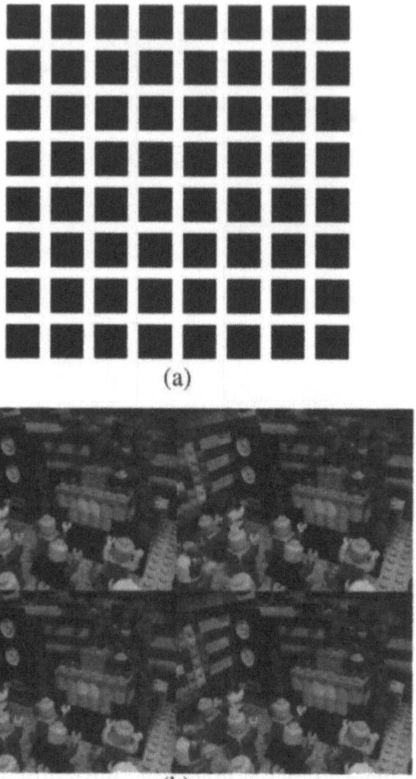

(a)

(b)

Figure 3.5: (a) The Tsai grid calibration image, (b) the "Lego" test image.

with the LCD panel, switching between calibration and test images. Why not simply extract Harris features from the Tsai grid images, use them to compute the ML homographies, and then assess registration accuracy against the ground-truth? The reason is simply that the Tsai images are very artificial, having only 4 types of feature point (at the corners of each black square). We might expect the localization error of the corresponding Harris features to vary rather systematically, possibly biasing the estimated homographies. The test images are intended to be fairly generic, whilst being quite rich in Harris features. They are a far more realistic approximation of the type of scenes that we might ask our feature-based algorithm to register in practice.

RMS point-transfer accuracy The RMS point-transfer error is evaluated over a 500×500 pixel region in the centre of each 720×576 pixel image. Figure 3.9 shows the transfer error vectors, evaluated using Equations (3.10a) and (3.10b), for two different image pairs. Note that the vectors are magnified 100 times. As expected, the direction and magnitude of the error varies slowly across the image, since small perturbations of the homography parameters away from the ground-truth result in structured changes in pixel transfer.

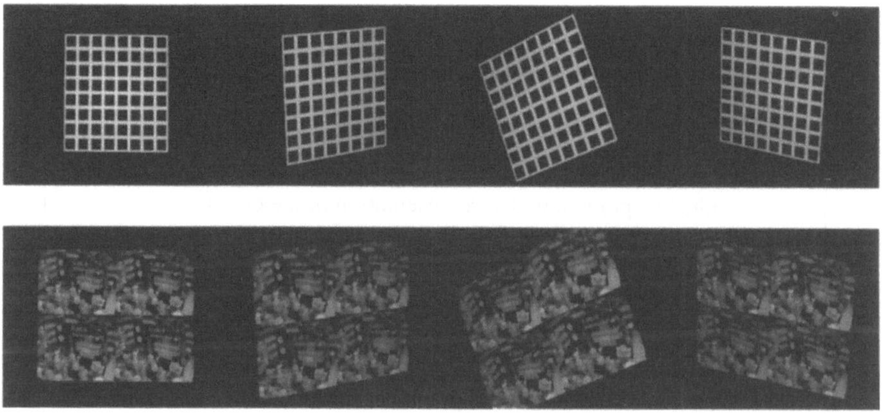

Figure 3.6: The captured images generated by viewing the LCD panel screen from four different view-points and remotely switching between the calibration image and the three test images. The Tsai grid and "Lego" images are shown. The images are related by full 8-dof projective transformations.

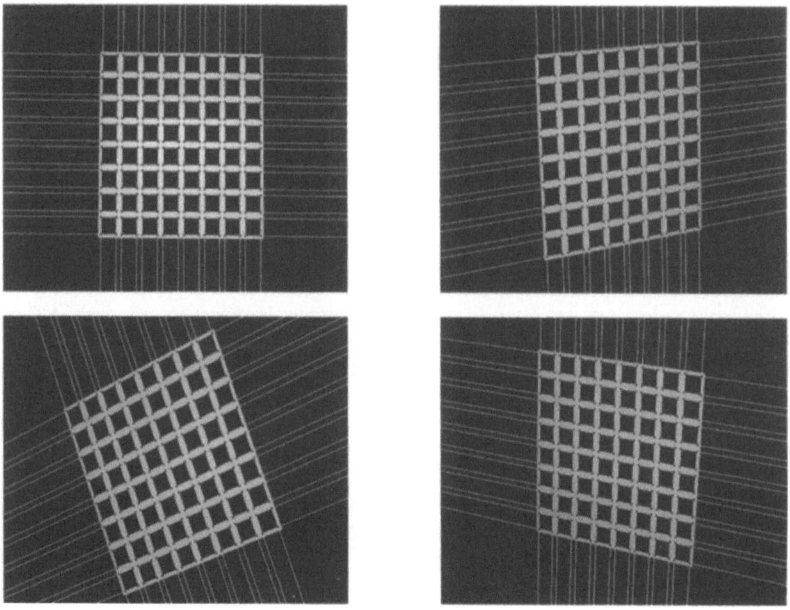

Figure 3.7: Line templates are fitted to Canny sub-pixel edges extracted from each view of the calibration image.

<u>Objective</u> Fit a line template to a view of the calibration image.

<u>Algorithm</u>

1. Process the image with Canny's sub-pixel edge detector [27]. This results in a point-sampled representation of the edges.
2. Transform the 16×16 line template under a putative homography H, which maps the template into the calibration image.
3. Measure the orthogonal distance of each edgel to each line in the template, and assign each sample to the closest line. Samples for which the distance is > 0.5 pixels are considered to be outliers.
4. Using the Levenberg–Marquardt algorithm, optimize the parameters of H so as to minimize the point-to-line residuals.
5. Repeat step 2 until convergence.

The initial estimate of H is determined by manually selecting the four corners of the calibration grid.

Table 3.2: The algorithm used to fit a line template to each view of the calibration image.

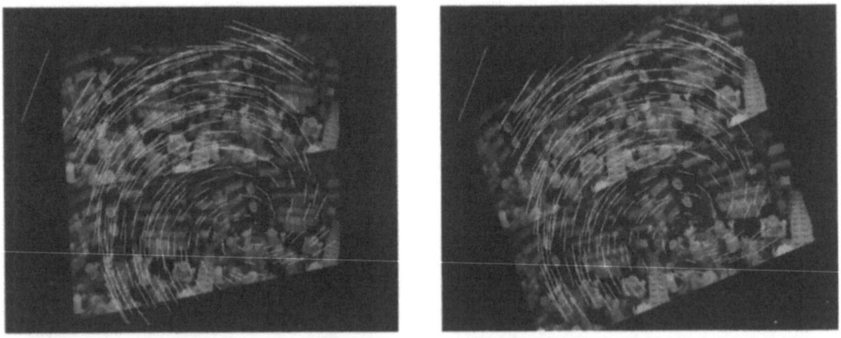

Figure 3.8: Approximately 300 feature correspondences between two images in the "Lego" set which are used to compute the maximum-likelihood homography.

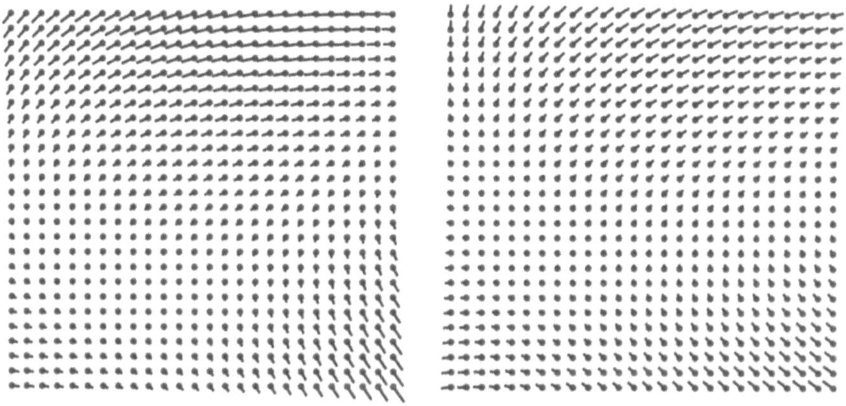

Figure 3.9: Forward and backward pixel-transfer error fields for the homography between images 1 and 2 in the "Mosaic" set, computed over a 500 × 500 pixel region in the centre of each image. Note that the vectors are magnified 100 times. The RMS error in this example is 0.10 pixels.

Image pair	Mosaic	Relativity	Lego
$H_{0\to1}$	0.060 (0.188)	0.068 (0.189)	0.075 (0.193)
$H_{0\to2}$	0.117 (0.211)	0.057 (0.176)	0.118 (0.207)
$H_{0\to3}$	0.102 (0.242)	0.092 (0.293)	0.110 (0.196)
$H_{1\to2}$	0.100 (0.222)	0.124 (0.378)	0.108 (0.258)
$H_{1\to3}$	0.104 (0.274)	0.143 (0.399)	0.141 (0.255)
$H_{2\to3}$	0.170 (0.400)	0.126 (0.265)	0.168 (0.215)

Table 3.3: The RMS and maximum (in brackets) pixel transfer errors compared to the ground-truth. The results indicate that our registration method is accurate to around 0.1 pixels. The maximum error is nowhere greater than 0.4 pixels.

For every pair of images in each test sequence, the RMS transfer error is computed using Equation (3.11). The results are tabulated in Table 3.3. The results indicate that our registration method is accurate to around 0.1 pixels.

Feature localization accuracy We now attempt to verify our broad assumptions about Harris feature localization: that they are distributed around the true, underlying point according to an isotropic, normal distribution with $\sigma < 0.25$ pixels.

A convenient statistic to use for this purpose is the distance between a (inhomogeneous) feature point $\mathbf{x}_2$ and its transferred corresponding point $\mathbf{x}'_1 = \mathtt{H}(\mathbf{x}_1)$,

$$\mathbf{d} = \mathbf{x}_2 - \mathtt{H}(\mathbf{x}_1)$$

Under the assumptions of isotropy and normality, each feature point has a 2 × 2 diagonal covariance matrix,

$$\Lambda_{x1} = \Lambda_{x2} = \sigma^2 \mathtt{I}$$

The covariance of point $\mathbf{x}_1'$ can be obtained using the Jacobian $\mathtt{J} = \frac{d\mathbf{x}_1'}{d\mathbf{x}_1}$,

$$\Lambda_{x1'} = \sigma^2 \mathtt{J}^\top \mathtt{J}$$

Note that, since we are dealing with homographies, the Jacobian J varies over the image. The covariance of the statistic d is then

$$\Lambda_d = \sigma^2 (\mathtt{I} + \mathtt{J}^\top \mathtt{J})$$

Finally, we can transform this distribution back to an isotropic one by means of a spatially varying affine transformation

$$\mathbf{d}' = \left(\sqrt{\mathtt{I} + \mathtt{J}^\top \mathtt{J}}\right)^{-1} \mathbf{d}$$

So for any set of feature correspondences, the method is to measure the vector $\mathbf{d}$ for every pair of corresponding features, and to map $\mathbf{d}$ to $\mathbf{d}'$ in the isotropic frame under the local affine transformation. If our assumptions are correct, we should end up with an isotropic, normal distribution with $\sigma_x = \sigma_y < 0.25$. To avoid any accusations involving the central limit theorem, we only apply this analysis to single image pairs, rather than combining the correspondences from all images into one large data set.

Figure 3.10 shows the distribution of the individual x and y components of the $\mathbf{d}'$ computed using corresponding feature-points in images 1 and 2 from the "Mosaic" and "Lego" test sets. Given our assumptions, we expect the two components to be independently and normally distributed with equal variances. The fitted normal distributions are overlaid on each histogram, from which the normality assumption seems reasonable. In fact, the normality may be verified using a χ^2 goodness-of-fit test [142, 160].

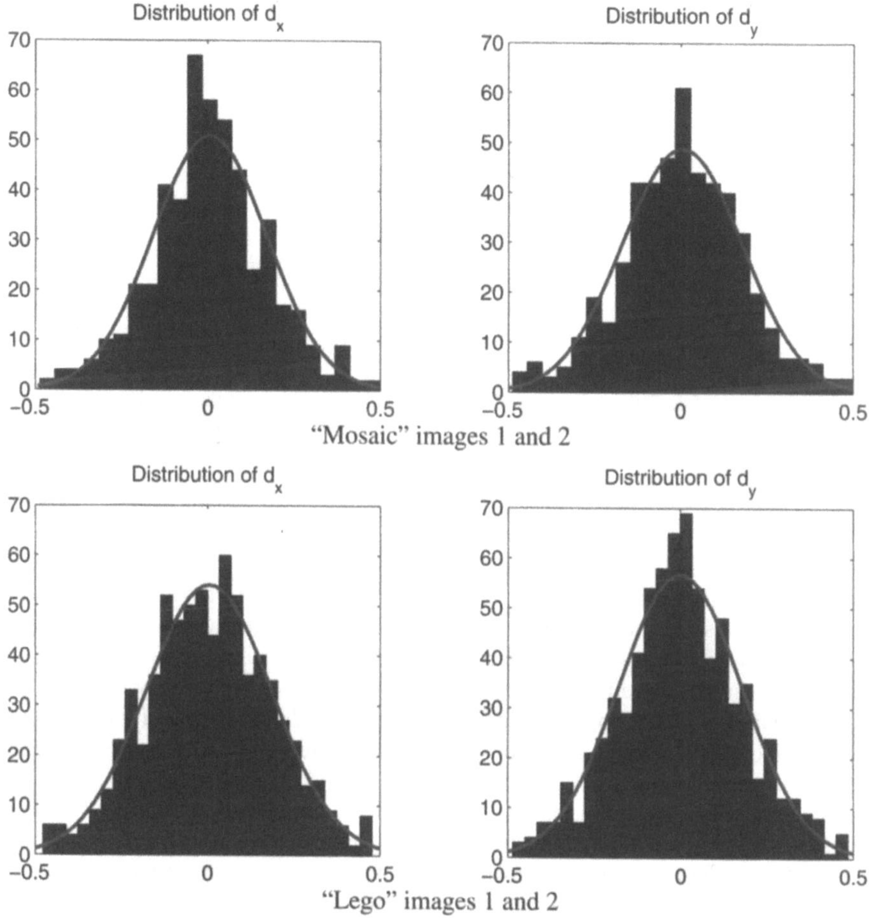

Figure 3.10: The distribution of the $\mathbf{d}'$ computed using corresponding feature-points in two different image pairs. The assumption that the individual x and y components are normally distributed with equal variances is supported by the fitted normal distributions which are overlaid.

Table 3.4 shows the standard deviations and covariances of the x and y components of the $\mathbf{d}'$ computed using the point correspondences between several pairs of images. There are three important points to note. First, σ_x and σ_y agree fairly closely for each pair of images. Second, the σ are all very small, less than 0.25 pixels. Third, the covariances are all tiny, indicating very little correlation between the d'_x and d'_y. Together, these three observations validate our assumptions of isotropy, normality and $\sigma < 0.25$ pixels.

Discussion The measured RMS transfer error of ~ 0.1 pixels between the ground-truth homographies and those computed using Harris features seems at first to be rather large. After all, the homographies were estimated by fitting just eight parameters

Image pair		σ_x	σ_y	σ_{xy}
Mosaic	$0 \to 1$	0.137	0.136	4.7×10^{-5}
	$1 \to 2$	0.166	0.174	1.1×10^{-3}
	$2 \to 3$	0.177	0.197	3.2×10^{-3}
Relativity	$0 \to 1$	0.138	0.143	2.4×10^{-3}
	$1 \to 2$	0.157	0.163	4.8×10^{-4}
	$1 \to 2$	0.152	0.185	1.3×10^{-3}
Lego	$0 \to 1$	0.153	0.126	1.3×10^{-3}
	$1 \to 2$	0.183	0.178	2.9×10^{-4}
	$1 \to 2$	0.176	0.205	6.5×10^{-3}

Table 3.4: Standard deviations (measured in pixels) and covariances of the x and y components of $\mathbf{d}'$ computed using the feature-point correspondences between several pairs of test images. As explained in the text, these results validate our assumptions about the localization of Harris feature points.

to several hundred point correspondences, and we have just demonstrated that the localization noise on those points really is normally distributed with $\sigma \approx 0.25$ pixels. So surely, the parameters of the homography should be estimated with a very high degree of accuracy? Well of course, they are; but that does not directly imply a tiny RMS transfer error. The probable blame lies with the *scale* and *perspective* parameters of the homography. Tiny perturbations to these parameters can cause large movements in the transferred points.

To relieve any remaining doubts, we can conduct an experiment using synthetic data. Sets of synthetic point correspondences are generated by transforming 300 randomly distributed points under known homographies. Isotropic, zero-mean Gaussian noise with $\sigma_x = \sigma_y = 0.25$ pixels is added to the point positions. The noisy correspondences are then used to compute ML homography estimates. Comparison of the estimated homographies with the ground-truth reveals exactly the same result as was observed in the real image experiment: an RMS transfer error of ~ 0.1 pixels.

Conclusion As stated at the start of this section, the purpose of this exercise was to get a feel for the accuracy of our registration method, and to get some idea as to whether the assumptions made are valid when applied to the kind of images that we expect to be dealing with. These goals have been achieved. But the reader is warned to regard the results with caution in the broader scope, since no mention has been made of image noise or blur, which if severe, can have a dramatic effect on feature localization.

3.6 Feature-based vs. direct methods

The principle behind direct intensity correlation methods has already been explained in Chapter 2 and will not be repeated here. To summarize, they seek a homography which maximizes a similarity measure between the images when one of them is warped according to the homography and compared with the other. In this section, the feature-based approach and direct methods are contrasted.

Invariance An important motivation for using features is their invariance to a wide range of photometric and geometric transformations of the image. The localisation of intensity discontinuities and auto-correlation maxima, the building blocks of many edge and point feature detectors, is unaffected by large illumination changes. Furthermore, the positions of these features map directly under projective transformations. This invariance is demonstrated empirically in the case of Harris features by Schmid *et al.* [130].

In direct methods there are a variety of ways by which to obtain photometric invariance. One may assume some model for the photometric mapping between the two images, for example that the intensities of corresponding pixels are related by a constant offset or an affine transformation. This requires extra parameters to be included in the registration to "soak-up" the photometric mapping. Alternatively, one might choose an image similarity metric which is invariant under the chosen photometric model, for example normalized cross-correlation, which is invariant to affine transformations of intensities. A third possibility is to pre-filter the images to remove the lowest frequency components prior to registration.

Direct methods are arguably at a disadvantage here. Any differences between the two images that are not accounted for by either the geometric transformation and photometric model must be absorbed by the noise model and similarity metric. If the photometric model is inadequate, corresponding pixels in the two images may exhibit large differences *even when the geometric registration is exact.* These artficially large residuals can dominate the similarity score and bias the final estimate of the geometric transformation.

Robustness to outliers Occlusions and specularities threaten the accuracy of any registration algorithm. In such cases, parts of the images will map correctly under a homography, whilst other parts (which have been obscured by other objects, shadows, etc.) will not. Consequently, different feature matches will be consistent with different motion models.

In feature-based methods, robustness to these outliers can be achieved simply and at low computational cost through the use of algorithms such as RANSAC (Random Sampling and Consensus) [153], which as demonstrated in Section 3.4, easily sift the dominant motion from correspondence sets heavily polluted with mismatches.

The use of random sampling methods in feature-based registration is feasible because of the low computational cost of scoring putative solutions. In direct methods this cost is much higher, requiring a full image warp and similarity score for each putative solution, making sampling methods much less attractive. However, alternative means of achieving robustness in direct methods have been proposed. One approach is to use a similarity metric under which the influence of large residuals is reduced, such as the M-estimator approach of Black and Anandan [16]. An alternative and closely related scheme is a multi-scale, iteratively re-weighted least-squares [80, 87] approach. However, both of these approaches are computationally expensive: optimizations involving non-convex M-estimators are generally difficult and are not guaranteed to converge to a global optimum. Sampling methods have an arguable advantage in this respect.

Applicability to general scenes One notable drawback to feature-based approaches is that the type and scale of features which are used for registration should ideally be chosen depending upon the type of scene in question. Point features, which are most commonly used, are suitable for the vast majority of everyday images, but for more specialist applications they do not provide the required information. Some examples might be images of smooth objects, scenes with low contrast, little edge information, or "pseudo-static" scenes, such as a lake or waving grass. In each case, there may still be an appropriate choice of feature type and scale which would result in a good feature-based estimator, but this choice requires significant scene understanding or user input. Furthermore, higher level features, such as curve segments, present extra difficulties when it comes to finding corresponding features across multiple images.

Partly due to their typical multi-scale, coarse-to-fine implementation, and partly because they do not distinguish between discontinuous and smoothly varying regions of the image, it can be argued that direct methods are more generally applicable to a wide range of scenes.

Computational efficiency The computational cost of feature-based method breaks down into feature detection (linear in the number of pixels), feature matching (at worst $O(N^2)$ in the number of features, but typically far less), and robust optimization (< 10 parameters, < 1000 residuals). The cost of a typical direct method is at every iteration, a couple of image gradient operations, an image warp per estimated parameter and a similarity score operation (all linear in the number of pixels).

The dominant cost in feature-based methods is that of determining putative feature correspondences. For small motions, the size of the search region can be small, and correspondence rapid. In wide-baseline applications, the number of putative matches that must be considered for each feature grows rapidly, as does the complexity of geometrically invariant metrics which must be used to score their similarity. The use of higher level image features, such as lines or curves, also greatly increases the complexity of the correspondence problem. However, once correspondence has been obtained, albeit polluted by outliers, the remaining optimization steps are extremely rapid.

The costs involved in direct methods may seem "bulky", but the use of coarse-to-fine strategies means that, particularly in small motion cases, a good optimization scheme converges in only a few iterations. Consequently, carefully implemented direct methods have proved just as successful in real-time tracking applications as their feature-based alternatives.

3.7 Photometric registration

This section describes a simple model of the global photometric differences which are often observed between two images of the same scene. A robust method for estimating the model parameters from geometrically registered input images is described. The estimated parameters may be used to remove the photometric difference between the images by altering the colour balance of one of them. By analogy with the geometric registration method previously described, we call this process *photometric registration*. The effectiveness of the scheme is demonstrated using three pairs of real images, each with a different source of photometric difference.

3.7.1 Sources of photometric difference

The image sequences used later in this book require a photometric model that can cope with two principal sources of global photometric difference, which can be summarized as follows.

- **Automatic camera adjustments** such as automatic gain control, white balance or exposure (as in Plate 5).
- **Illumination change** due to variation in the ambient light level or daylight level, or due to relative motion between the scene and the light source (as in Plate 6).

A slightly more unusual situation in which we apply photometric registration is illustrated in Plate 3, which shows images of two physically different scenes. At the structural level, the images are almost identical (modulo a geometric transformation), but there are severe photometric differences between them.

3.7.2 The photometric model

The model treats each of the red, green and blue colour channels independently. Within each channel, the variation between the two images is modelled as a linear transformation, having two parameters: a multiplicative term α, and an additive term β. Expressing the image pixels as 3-element vectors, the transformation can be written as

$$\begin{pmatrix} r_2 \\ g_2 \\ b_2 \end{pmatrix} = \begin{bmatrix} \alpha_r & 0 & 0 \\ 0 & \alpha_g & 0 \\ 0 & 0 & \alpha_b \end{bmatrix} \begin{pmatrix} r_1 \\ g_1 \\ b_1 \end{pmatrix} + \begin{pmatrix} \beta_r \\ \beta_g \\ \beta_b \end{pmatrix} \tag{3.12}$$

requiring six parameters in total. This simple model proves to be rich enough for the purposes described here. There is no apparent benefit in using a full affine transformation of the RGB colour-space.

3.7.3 Estimating the parameters

The procedure for estimating the six photometric parameters requires the two images to first be accurately registered, using the method described in Section 3.4, and warped into alignment. The remaining differences between corresponding pixels in the aligned images can then hopefully be "absorbed" by the photometric model. Treating each channel separately, the estimation procedure for α and β is clearly a simple line-fit to the intensities of corresponding pixels (i_1, i_2), easily achieved by orthogonal regression.

However, at many pixels, the difference in intensities, $i_1 - i_2$, cannot be explained by the linear photometric model. These model-outliers arise from various sources: saturation at the low or high end of the pixel intensity range, specularities or occlusions, shadows. Just as in the geometric estimation algorithm, it is vital that the line-fitting algorithm be robust to these model-outliers.

The algorithm used for this purpose is Torr and Zisserman's MSAC (M-estimator SAmple Consensus) algorithm [159], a variation of the previously described RANSAC algorithm. Evidence suggests that this algorithm gives better results than the RANSAC algorithm in problems where the proportion of outliers is high, as is the case here. The algorithms differ only in the way in which the quality of a proposed solution is evaluated. Instead of counting the number of inliers, as RANSAC does, MSAC determines the likelihood of the data given the proposed model parameters according to a simple, robust M-estimator [77, 78]. For each data point $(i_1, i_2)_n$, the distance d_n to the proposed line (α, β) is computed, and the overall cost associated with the solution is given by

$$C = \sum_{\forall n} \rho(d_n^2) \tag{3.13}$$

where

$$\begin{aligned} \rho(d^2) &= d^2 \ \text{ if } d^2 < T^2 \\ &= T^2 \ \text{ otherwise.} \end{aligned} \tag{3.14}$$

The threshold T is set to 1.96σ so that Gaussian inliers are only incorrectly classified as outliers 5% of the time, and we assume σ is around 5.0 "grey-levels".

After evaluating many putative solutions, the MSAC algorithm returns the parameters (α, β) with the lowest score. In the final step, the parameters are refined by performing orthogonal regression on the MSAC inliers.

3.7.4 Results

We now apply the photometric registration algorithm to the three examples shown in Plates 3, 5 and 6. In each case, having estimated the six photometric parameters, the first image in each pair is "corrected" by applying the appropriate linear transformation to each colour channel.

Bank-note pair This example shows two different specimens of the same denomination of currency, captured using a flat-bed scanner. Image 1 has been treated with a purple dye which stains oily residues, such as fingerprints. Unfortunately, low-quality bank-note papers, such as US dollars, also absorb some of the dye, taking on a pink colour. Plate 3 shows the geometrically registered images (only the mutually overlapping portion is shown), and below, density plots of corresponding pixel intensities

(i_1, i_2) in each colour channel. The robust MSAC line-fit is overlaid on each graph. Plate 4 shows the original and corrected images and intensity-profile graphs. The corrected image has lost the pink hue, and a purple latent fingerprint mark is just visible on the President's face. In this case, the red and green channels required the most severe correction. The profiles of the corrected image match closely those of image 2.

Indoor pair This example is taken from sequence of images captured by a rotating camera. The photometric difference between the frames shown is caused by the camera's automatic white-balance and gain control. Plate 5 shows the original and corrected versions of image 1 on either side of image 2. Also shown are intensity-profiles along a slice taken through the images. The bottom three graphs show the original (red) and corrected (green) profiles from image 1 overlaid on the profiles from image 2. The corrected profiles match the desired profiles very closely.

Museum pair In this example the difference is due to a break in cloud cover, meaning that the building is bathed in sunlight in image 2. Consequently, there are many outliers at pixels which are in shadow in image 2. Plate 6 shows the original and corrected images and intensity-profile graphs. The corrected image has taken on the orange glow exhibited in the sun-lit image. In this case, the red channel required the most severe correction. Again, the profiles of the corrected image match closely those of image 2.

3.8 Application: Recovering latent marks in forensic images

In this section, we describe an application which uses the above registration techniques to allow latent marks to be recovered from forensic images. The aim is the removal of distracting background patterns from forensic evidence so that the evidence is rendered more visible. An example is the image of a fingerprint on a non-periodic background. The method involves geometrically and photometrically registering the image with a control image of the background pattern that we seek to remove. Image differencing then reveals the latent mark. The method applies in situations where *periodic background* removal, accomplished by Fourier transform techniques [22, 66, 150, 170], would not be successful. The work is described in detail in [30] and patent GB2342207A. All the example images in this section were supplied by the UK Forensic Science Service, and the research was carried out as part of the EU project "Image Processing for Forensic Science Support" (IMPROOFS) [3].

3.8.1 Motivation

Marks of interest to a forensic examiner are often encountered overlaid on a background pattern. In many instances this background pattern can hinder or even prevent the reading of the detail of interest in the mark. An example is shown in Plate 7a. Thus, the requirement of a forensic examiner is to have a tool that attenuates the unwanted background signal whilst leaving the mark detail of interest unaffected. One of the most widespread tools is the use of Fourier domain filtering methods for the attenuation of periodic background patterns interfering with the reading of a fingerprint pattern [22, 23, 66, 150, 170]. The major drawback to Fourier filtering is the requirement that the pattern is repetitive across the area of interest. In many instances this is not the case, for example most bank notes have very irregular patterns designed into them. These non-periodic patterns do not produce a regular pattern of spikes in the Fourier domain, and hence conventional background attenuation methods cannot be applied.

3.8.2 Method

The basic method is as follows: given an image containing a mark which is camouflaged by the background pattern, and another image of the same scene, not necessarily aligned with the first image, but in which the mark is not present, produce an image of the mark separated from the background, thereby rendering it more visible. The image containing the latent mark is referred to as the *evidence image*, and the clean image as the *control image*. As an example, we shall consider the two images shown in Plate 7a and 7b, which show a section of a British five-pound note, scanned at 600 dpi on a flat-bed scanner.

The method assumes that any differences between corresponding pixels in the evidence and control images will be due to the presence of the latent mark occluding pixels in the evidence image. In order for this assumption to be true, the images must be aligned to sub-pixel accuracy, and their global photometric characteristics must be the same. However, in practice the evidence and control images will be taken at different times, possibly using different equipment with different settings, hence they will neither be geometrically aligned nor colour balanced. To compensate for this, we make use of the registration methods described in Sections 3.4 and 3.7 in order to bring the control image into accurate alignment with the evidence image, both geometrically and photometrically. The steps are as follows.

1. **Geometric registration** Compute Harris interest point features in both images, and register them using the algorithm of Section 3.4.

2. **Photometric registration** Using the computed registration parameters, warp the control image into alignment with the evidence image, and apply the robust photometric registration method described in Section 3.7.

3. **Attenuation** Using the estimated photometric parameters, transform the colour channels of the control image. The two images should now be exactly alike modulo the latent mark. Subtract the control image from the evidence image to obtain the separated image.

Often, such as is the case with bank notes imaged using a flat-bed scanner, a full 8 dof homography is unnecessary for aligning the two images. In these cases, greater accuracy can be achieved by modifying the geometric registration algorithm to compute a lower order transformation, such as a 6 dof affine transformation or 4 dof similarity transformation.

Plate 7c shows the five-pound note control image after geometric and photometric registration. The control image is aligned and colour balanced to match the evidence image. The difference image, formed by subtracting the control from the evidence, is shown in Plate 7d. The image has been intensity normalized and inverted to enhance the mark. The fingerprint is now clearly visible.

This example also exhibits a frequent problem when working with bank note images - some of the text is still visible. This is an artifact of the printing process: notes are generally composed from several printed layers, and the relative displacements of these layers can vary from one batch of notes to another by several pixels. To compensate for this would require several independent homographies to be computed between the images, one for each layer. The single homography fitted by the RANSAC algorithm only accounts for the dominant background layer, and consequently the text in this example is not correctly aligned.

It is important to note that this method relies very strongly on the use of robust algorithms. Clearly, a great many corresponding pixels in the two images have very different values, whether because they are occluded by the latent mark, or because they exist in a different screen printing layer. It is imperative that the geometric and photometric alignment is robust to these outliers. The feature-based RANSAC algorithm is ideally suited to this task.

evidence *registered control*

Figure 3.11: The US dollar evidence and control images after geometric and photometric alignment. As is evident in Plate 3, the photometric registration phase is extremely important in this example.

3.8.3 Further examples

Figures 3.11 and 3.12 show the stages of processing for the US dollar bill images first seen in Plate 3. In this example, the fingerprint is completely invisible in the evidence image. The low-quality paper on which these bills are printed absorbs some of the purple disclosure dye. The photometric registration is able to compensate for this, as demonstrated in Figure 3.11. The final result, Figure 3.12, quite clearly shows a partial print in the centre of the image.

Figures 3.13, 3.14 and 3.15 show the stages of processing for two images of a PVC floor covering with a highly irregular patterning. The evidence image contains two partial marks made by the tread of a shoe, although this is difficult to see in the original image. The difference image, shown in Figure 3.15, clearly shows the two marks.

Figure 3.12: Although almost impossible to see in the evidence image, the difference image clearly shows a partial fingerprint.

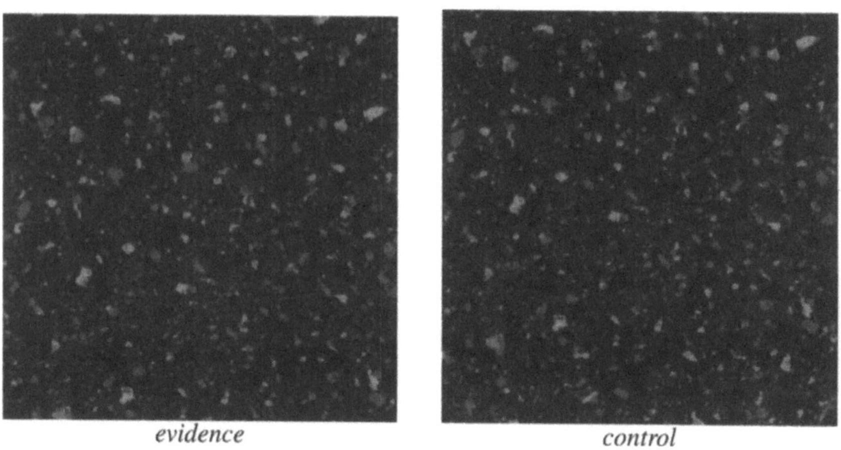

Figure 3.13: Images of a plastic floor covering captured with a digital camera. The evidence image contains a partial shoe-mark, although it is extremely difficult to see.

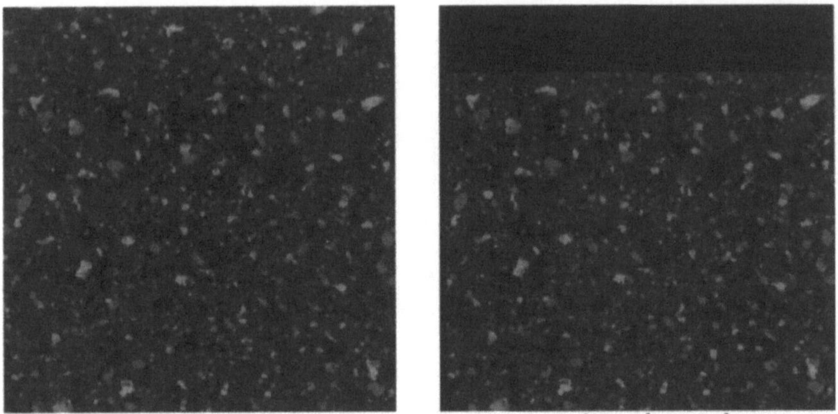

evidence *registered control*

Figure 3.14: After registration, the control image is aligned with the evidence image to sub-pixel accuracy.

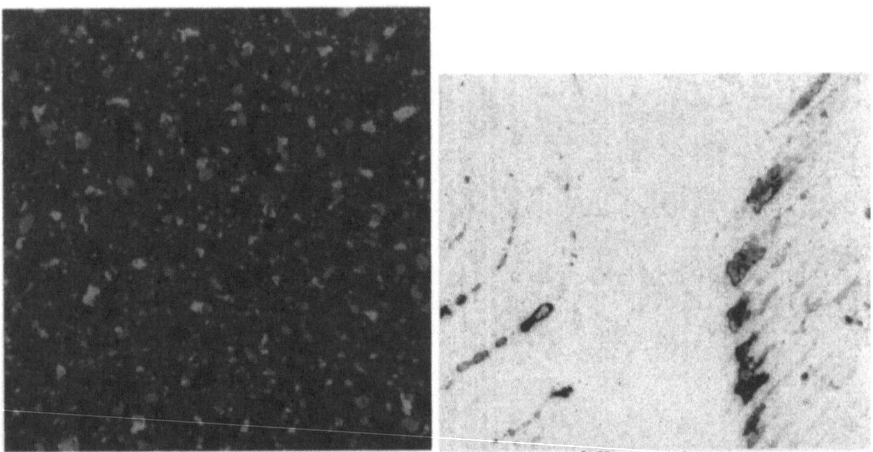

evidence *difference image*

Figure 3.15: Although very hard to see in the evidence image, the difference image clearly shows two partial marks made by the tread of a shoe.

3.9 Summary

In this chapter, the problems of accurate geometric and photometric registration of images has been considered. An automatic algorithm has been described for the registration of images under the assumption of a projective motion model using corresponding feature points in two images. The algorithm has been shown to be highly robust to mismatched features, whilst still being computationally very efficient. Through empirical investigation, the accuracy of the method was shown to be very high. This will prove suitable for use in the super-resolution algorithms described in later chapters.

A simple model was proposed to account for the photometric differences between pairs of images which may be caused by illumination changes, camera gain, or white-balance. A robust and efficient algorithm for estimating the model parameters given geometrically registered images was described, and demonstrated to be effective in compensating for photometric differences between a variety of real images.

Finally, a forensic application which utilizes both geometric and photometric registration was described and its effectiveness demonstrated with real examples.

Summary

Chapter 4

Image Mosaicing

4.1 Introduction

Image mosaicing is the alignment of multiple images into larger compositions which represent portions of a 3D scene. The mosaicing methods described in this chapter are concerned with images which can be registered by means of a planar homography: views of a planar scene from arbitrary viewpoints, or views of a scene taken by a rotating camera. There are three important factors to consider when constructing a mosaic: the estimation of a set of homographies which are consistent over all the views; the choice of reprojection manifold on which the images are composited; and the choice of algorithm for blending the overlapping images.

Section 4.2 outlines the basic steps in constructing a mosaic representation of a scene, and in rendering an image from the mosaic. Section 4.3 details the rendering process, describing the most commonly used reprojection manifolds and blending methods, and demonstrating how the photometric registration method described in Chapter 3 is effective at eliminating visible "seams" between different source images in the rendered composites. Section 4.4 describes how the two-view ML homography estimator of Chapter 3 is extended to perform simultaneous, globally consistent registration of N-views. To facilitate the use of this estimator, we describe a novel and efficient algorithm for matching feature points over many views. Section 4.5 is concerned with the specific case of the *planar* reprojection manifold. We describe a new algorithm for automatically orienting the manifold so as to produce a well-balanced and aesthetically pleasing rendered composite. Finally, Section 4.6 comments briefly on the many possible applications of image mosaicing proposed in the literature.

4.2 Basic method

The construction of image mosaics is described in the literature in several contexts, but for clarity we briefly review the basic method.

4.2.1 Outline

There are three basic steps: registration, reprojection and blending, which are now described in more detail.

1. **Registration** The objective is to register every image into a global coordinate frame which contains the whole scene. The choice of global frame is unimportant, but for convenience it is usually chosen to be axis-aligned with one of the N input images, called the *reference image*, so that the transformation between the global frame and the reference image is simply the identity. For a given sequence of images, registration proceeds by first computing the geometric transformation between consecutive pairs of images, using a method such as that described in Chapter 3. The transformation between a particular image m_k and the reference image m_{ref} is then obtained by concatenating (multiplying) the intermediate sequence of homographies, for example

$$\mathtt{H}_{(\text{ref}\rightarrow k)} = \prod_{n=\text{ref}}^{k-1} \mathtt{H}_{(n\rightarrow n+1)}.$$

   However, this computation is prone to "dead-reckoning" error accumulation, which is a particular problem in image sequences which "loop back", imaging some part of the scene more than once at temporally separated points in the sequence. This issue is addressed in Section 4.4, where it is shown that a globally consistent set of homographies may be computed by simultaneously registering all views in the global frame.

2. **Reprojection** After registration, every point in every image can be transformed to a point in the global frame. The set of all images and homographies comprises the *mosaic representation* of the scene. In order to actually render an image from the mosaic representation it is necessary to supply a further transformation which maps points in the global frame to points in the rendered image. In some instances, this mapping may be as simple as a similarity transform (scaling and translation), or as discussed in Section 4.3, may be a more complicated transformation, such as a synthetic camera rotation or cylindrical polar mapping. Having chosen a rendering transformation, each image is then warped into the frame of the rendered image.

3. **Blending** The final stage in rendering from the mosaic is to blend the images together in their overlapping portions. As we have already seen in Chapter 3, it is often the case that significant global photometric differences can occur between images in a sequence. If not corrected, this can give rise to unsightly seams in rendered mosaics. The image blending function can be chosen to ameliorate this effect: common methods include simple averaging of intensity values, feathering and temporal median filtering. Better results may be obtained however, if photometric registration and correction is performed prior to rendering the mosaic. This problem is examined in more detail in Section 4.3.3.

The process is outlined schematically in Figure 4.1.

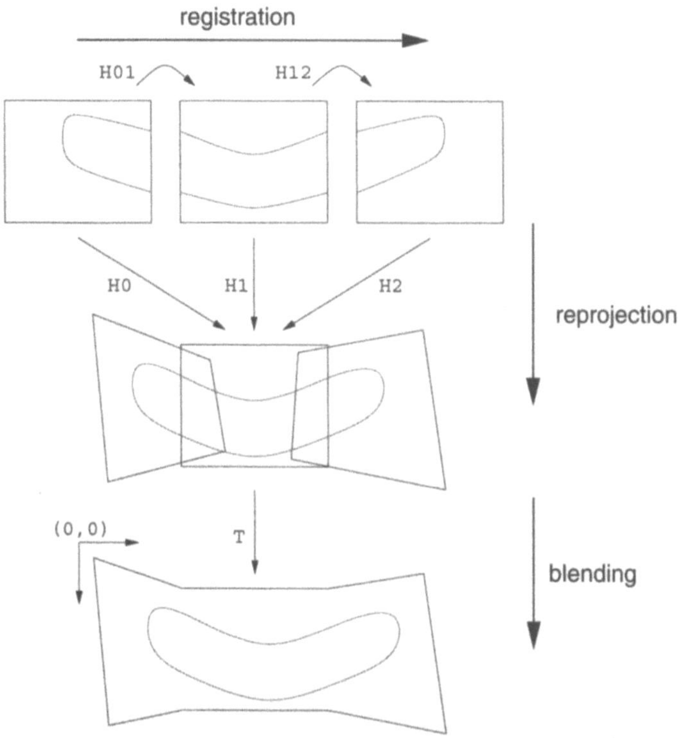

Figure 4.1: The three basic steps in forming a mosaic representation of an image sequence and rendering a novel view. In this example, the middle frame is being used as the reference frame, and hence H_1 is the identity. The rendering transformation T simply shifts the origin.

4.2.2 Practical considerations

Radial distortion Radial lens distortion is common in inexpensive camera optics and this can cause problems when mosaicing sequences captured using camcorders or cheap digital cameras. Szeliski *et al.* [145, 147] propose that the images be weighted heavily toward the centre when combining them in order to ameliorate the problem. A better solution is to remove the radial distortion from the images prior to mosaicing, using one of the methods already discussed in Chapter 3. An example of a mosaic created using this correction is shown in Figure 4.2.

Independent motion Suppose the camera rotates about its centre but during the acquisition there is movement in the scene, for example a person walking. The moving object will appear multiple times in the mosaic and, depending on the blending method, may look blurred and ghost-like, or may be chopped or truncated in some way. It is often desirable to remove these artifacts altogether. One approach is to use a temporal-median filter when combining the images. This removes the effect of temporary occlusions provided that the true background is unoccluded in more than half the frames. Irani *et al.* [85, 86] describe a more sophisticated approach in which

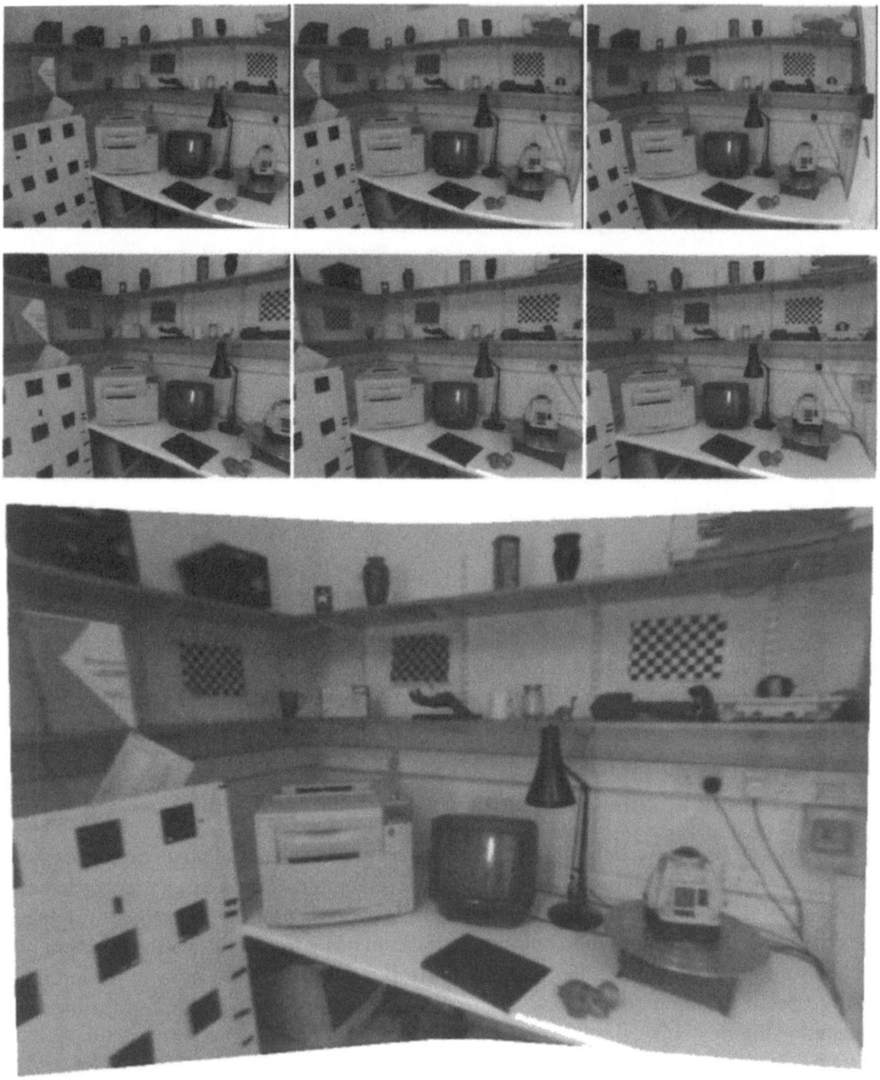

Figure 4.2: (Top) Three images captured by a rotating camera fitted with a very wide-angle lens. The images have severe radial lens distortion. (Middle) The images after radial lens correction. (Bottom) A mosaic image composed from the corrected images.

the multiple independently moving image regions are segmented and tracked through the sequence, starting with the dominant frame-to-frame motion. The algorithm produces high-quality representations of each region. A different approach is proposed by Davis [45], in which the mosaic image is segmented into several disjoint regions such that no moving object crosses the boundary of any region. The mosaic image is then assembled from the jigsaw-like segmented regions.

4.3 Rendering from the mosaic

The mosaic representation provides a one-to-one mapping between points in each image and some global coordinate frame. There are many ways in which this representation may be used, as discussed in section 4.6, but by far the most common requirement is to render a novel view of the scene. In this case, there are two factors which determine the appearance of the rendered image: the *reprojection manifold* and the *blending function*. We now discuss these concepts and examine some commonly occurring cases.

4.3.1 The reprojection manifold

To render from the mosaic, we must specify a transformation T which maps points in the rendered image to points in the global coordinate frame. T is referred to as the *rendering transformation*. In principle, T could be any one-to-one mapping, but in practice it is usually determined by the choice of *reprojection manifold*, the surface which plays the role of the imaging sensor in the virtual camera, for example a plane or a developable surface such as a cylinder. Points in the global frame are back-projected onto the manifold, and if the manifold is developable, it is simply "unrolled" to form the rendered image.

Planar projection The simplest and most commonly used manifold is a plane, onto which all of the images are reprojected, as shown in Figure 4.3. In this case, the rendering transformation is a homography. The rendered image has the classic "bow-tie" form since the projective distortion of the back-projected images increases toward the periphery of the mosaic. An example is shown in Figure 4.6(a). Note that straight lines are preserved. We can think of this as simulating a camera with an extremely large sensor array. In reality, of course, it would be very hard to construct such a camera: a large sensor array would be subject to *vignetting* [8] (shading) at the periphery. Planar projection mosaicing allows such images to be generated without these artifacts and without requiring exotic hardware. However, for sequences which sweep a large angle (> 90 degrees), the projective distortion means that the mosaic image becomes infinite in size, and the planar manifold cannot be used.

It is important to note that the planar manifold is suitable for *both* "general-scene,rotating-camera" *and* "planar-scene,general-motion" cases.

Cylindrical projection A cylindrical manifold, concentric with the rotating camera, is a suitable manifold on which to reproject rotating camera sequences which sweep

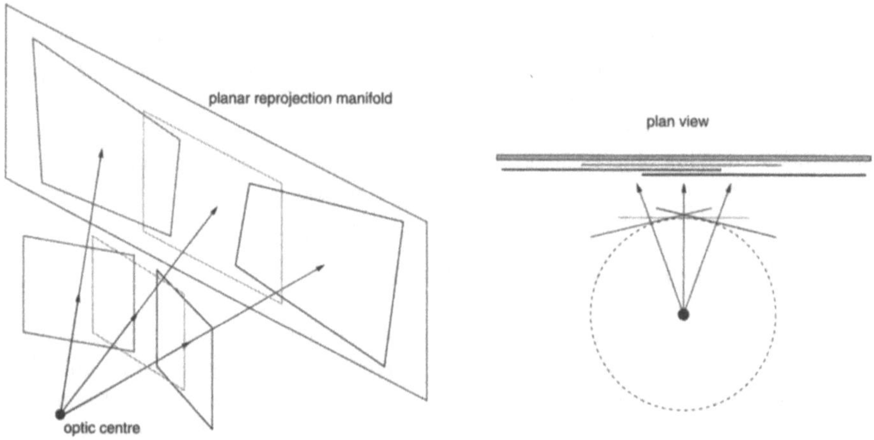

Figure 4.3: Schematic illustration of mosaic rendering by reprojection onto a planar manifold.

a very large angle (possibly a full 360 degree sweep), as shown in Figure 4.4. The method is best suited to image sequences in which the camera rotates about a single axis, in which case the input images are tangent planes to the manifold. The rendered image does not suffer from the same projective distortion seen in the planar manifold projections. Instead, straight lines in the world are mapped to sinusoids, as can be seen in Figure 4.6(b). The rendering transformation maps cylindrical polar coordinates to rectangular image coordinates as follows. A point (X, Y, Z) in the camera-centred coordinate system maps to a point (θ, v) on the manifold as

$$\theta = \tan^{-1}\left(\frac{X}{Z}\right) \qquad\qquad v = \frac{Y}{\sqrt{X^2 + Z^2}} \tag{4.1}$$

When the camera motion is pure rotation, a homogeneous 2D point $\mathbf{x} = (x, y, 1)$ in image n projects to a ray $\mathbf{X} = (X, Y, Z)$ in camera coordinates as

$$\mathbf{X} = \mathtt{R}_n^{-1}\mathtt{K}_n^{-1}\mathbf{x} \tag{4.2}$$

where $\mathtt{K}_n$ is the calibration matrix for the nth camera, and $\mathtt{R}_n$ is the rotation of the nth camera relative to the reference view. These equations are combined to obtain the mapping between image points (x, y) and points on the manifold (θ, v).

Evidently, this method requires an estimate of the camera calibration matrices $\mathtt{K}_n$. A non-linear method for accurately estimating the internal parameters of a rotating camera is described by Hartley [74, 75]. However, it is not essential for the calibration to be supremely accurate in order for this rendering technique to be effective for interactive display. Standard assumptions about the camera parameters (zero skew, square pixels, and principal point at the image centre), together with a ball-park estimate of the focal length, are perfectly adequate. With these assumptions, the much simpler, linear calibration method described by de Agapito *et al.* [46] provides a suitable estimate of the focal length.

As a further example of the application of this method to very wide angle sweeps, Figure 4.22 shows a full 360° panoramic mosaic projected onto a cylindrical manifold.

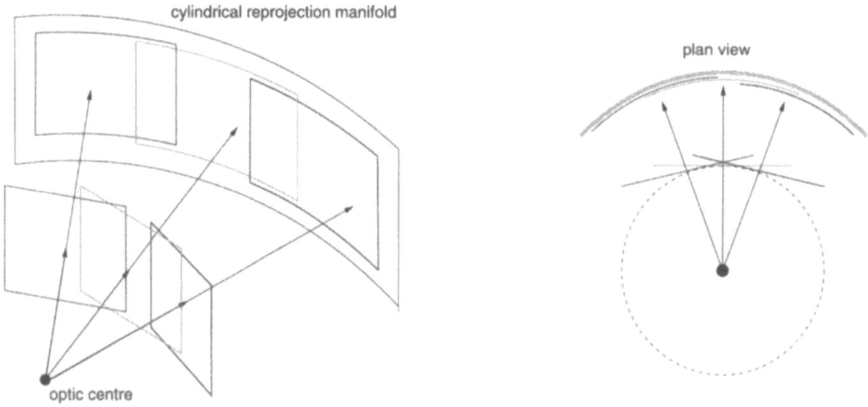

Figure 4.4: Schematic illustration of mosaic rendering by reprojection onto a calibrated cylindrical manifold.

Spherical manifolds In a similar scheme to the cylindrical manifold, points in the mosaic representation may be parameterized in terms of spherical polar coordinates. However, due to the obvious problems with singularities, and the lack of any satisfactory mapping from spherical to cartesian coordinates, there is little point in explicitly reprojecting the images onto such a manifold. A spherical manifold cannot therefore play the same role as the planar or cylindrical manifolds do in determining the rendering transformation.

That said, there are many references made to spherical panoramic representations in the literature: in image-based rendering [100], wide-baseline 3D scene reconstruction [89, 90] and virtual-reality applications [146]. In each case, image sequences are used which cover part or all of the view-sphere, and the mosaic simply provides a convenient *implicit* representation of the relative placement and orientation of each image on the view-sphere.

Rendering from such a mosaic is usually done by reprojecting a small portion of the view-sphere onto a planar manifold. This is the basis of the QuickTime VR software [35], and is discussed further in section 4.5. Another possibility is a Mercator projection [146]. Occasionally, to aid in visualization, an explicit spherical reprojection is performed by texture-mapping the images onto a 3D VRML model, which may be manipulated by the viewer [37, 38, 148].

Manifold projection Peleg's manifold projection method [108, 111] attempts to display the mosaic projected onto the *envelope* of the image planes from which it was acquired. The method forms the basis of Peleg and Herman's "VideoBrush" software [112]. The implementation details will not be repeated here, but essentially the method lies somewhere between planar mosaicing and calibrated projection onto a cylindrical manifold, and is closely related to the linear push-broom camera. The images are registered using only Euclidean transformations, and combined using the nearest image-centre method (see section 4.3.2). The rendered image is locally similar to the input images and of finite extent, without the need for calibration. It has

Figure 4.5: A sequence of 24 images captured by a rotating camera.

also been shown to generate reasonable mosaics when the camera motion constraints are violated. An example of a mosaic generated by manifold projection is shown in Figure 4.6(c).

4.3.2 The blending function

Each pixel in the rendered image corresponds to some pre-image point in the scene. Furthermore, each pre-image point is typically observed in several of the overlapping views. Consequently, when choosing the value of a pixel in the rendered image, it is possible to draw on information from several input images. The rendering algorithm for each pixel in the rendered image has the following basic form.

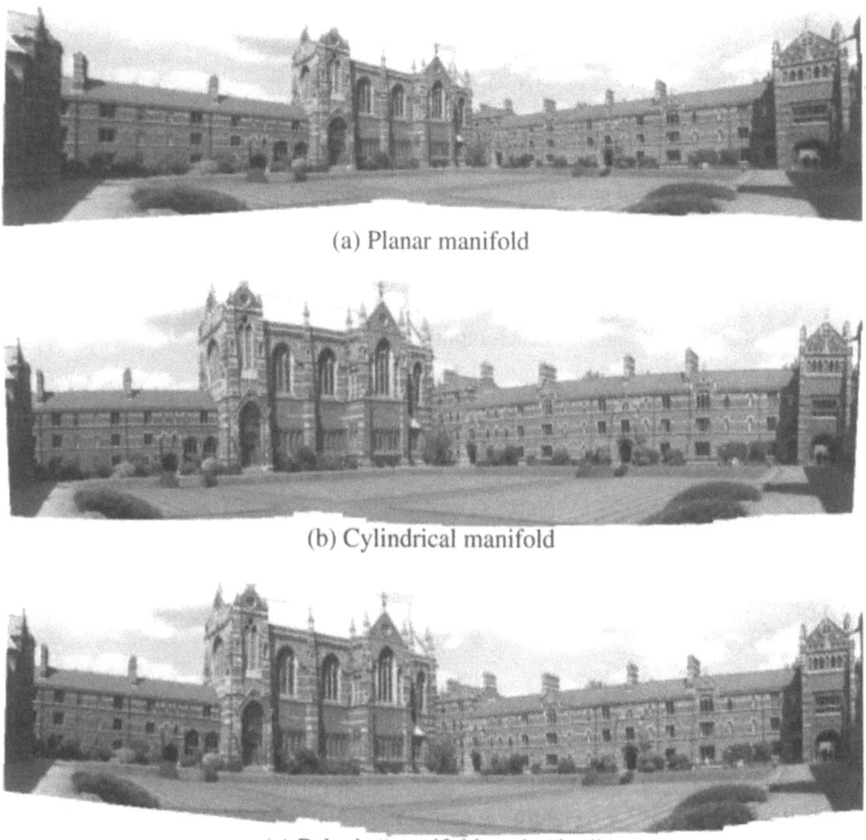

(a) Planar manifold

(b) Cylindrical manifold

(c) Peleg's "manifold projection"

Figure 4.6: Mosaics created from the image sequence shown in Figure 4.5. (a) Reprojection onto a planar manifold. Straight lines are preserved, but projective distortion increases toward the periphery. (b) Reprojection onto a cylindrical manifold. Straight lines are mapped to sinusoids, but there is no projective distortion, allowing large-angular sweeps to be rendered as a single image. (c) The mosaic created using Peleg's "manifold projection" method, in which the images are registered using only Euclidean transformations. The result is very similar to the cylindrical projection.

Figure 4.7: (Left) The bi-quadratic weighting function used to "feather" blended images. (Right) The function is used to set the transparency of an image.

1. Using the rendering transformation and pre-computed homographies, transform the pixel coordinates into each input image in turn.

2. For each image in which the transformed pixel lies inside the image boundary, sample the intensity value using a suitable interpolation scheme (e.g. bilinear, bicubic or area-sampling).

3. Combine the set of intensity values obtained into a single value using the chosen blending function.

We now describe a few common blending functions, examples of which are shown in Plate 8.

Averaging The output pixel value is obtained by simply averaging the values extracted from the input images.

Feathered blending A weighting function is associated with the input images, decaying from a maximum at the centre of the image, to zero at the image boundary. This function is used to weight the values extracted from the images when computing their average. An example weighting function is shown in Figure 4.7, in this case, a bi-quadratic function :

$$f(x, y) = (1 - (x - \tfrac{w}{2})^2)(1 - (y - \tfrac{h}{2})^2)$$

Nearest image centre When extracting values from the input images, the distance of the sampling location from the image centre is also computed. The set of values are ranked according to this distance, and the candidate closest to its image centre is taken as the output pixel value.

Temporal median filtering The rendered pixel value is the medioid of the values extracted from the images. The mediod is a robust estimator of the centroid. Consequently, mosaics composed using this blending method are robust to outliers caused by independent moving objects, specularities, etc.

4.3.3 Eliminating seams by photometric registration

In this section, we apply the photometric registration algorithm described in Chapter 3 to eliminate "seams" in sequences which are affected by serious photometric variations. In section 3.7, we described a robust method for performing photometric registration and correction of a pair of images by estimating the six parameters of a simple colour-space transformation. To apply this method to mosaicing, we first compute the photometric parameters relating consecutive pairs of images in the sequence. As with the geometric registration parameters, the computed colour-space transformations are easily chained together to compute the accumulated transformation between non-consecutive image pairs. One frame in the sequence is chosen as a reference, and every other image is photometrically registered with the reference, and corrected accordingly.

Plate 9 shows a mosaic rendered from a sequence of eight images taken by a rotating camera. The mosaic is composed using the nearest image-centre blending method. The camera's automatic white balance control has varied during the capture, causing considerable photometric differences between frames. Consequently, seams are clearly visible between the different images. Plate 9 also shows the same mosaic formed from the photometrically corrected images, using the same blending method. Frame 5 was used as the photometric reference frame. The seams have been eliminated.

4.3.4 Eliminating seams due to vignetting

Figure 4.8 shows a set of six aeriel images of the Niger delta region. The images show significant shading toward the periphery, an effect which is probably caused by *vignetting* [8]. Mosaics rendered from these images show severe artifacts due to the shading : noticeable seams if simple averaging is used; smooth intensity variations resulting in a "tiger-stripes" pattern if feathered blending is used.

Fortunately, it is fairly straightforward to correct this problem. Vignetting is modelled by a $\cos^4(\alpha)$ fall-off in intensity away from the principal point, but for simplicity we approximate this with a quadratic curve. The method applied here assumes that the optic axis passes through the centre of the image. The intensity fall-off at a radius r from the principal point is computed by taking the median of all the pixels in an annulus between radius r and $r + \delta r$. The median intensity is sampled at 100 increasing radii, and a quadratic curve fitted through the values obtained. The correction factor for a pixel at radius r is then $\frac{f(r)}{f(0)}$.

Figure 4.9 shows the corrected images, and the mosaic composed from them. The mosaic is now seam-free and the brightness is constant across the scene.

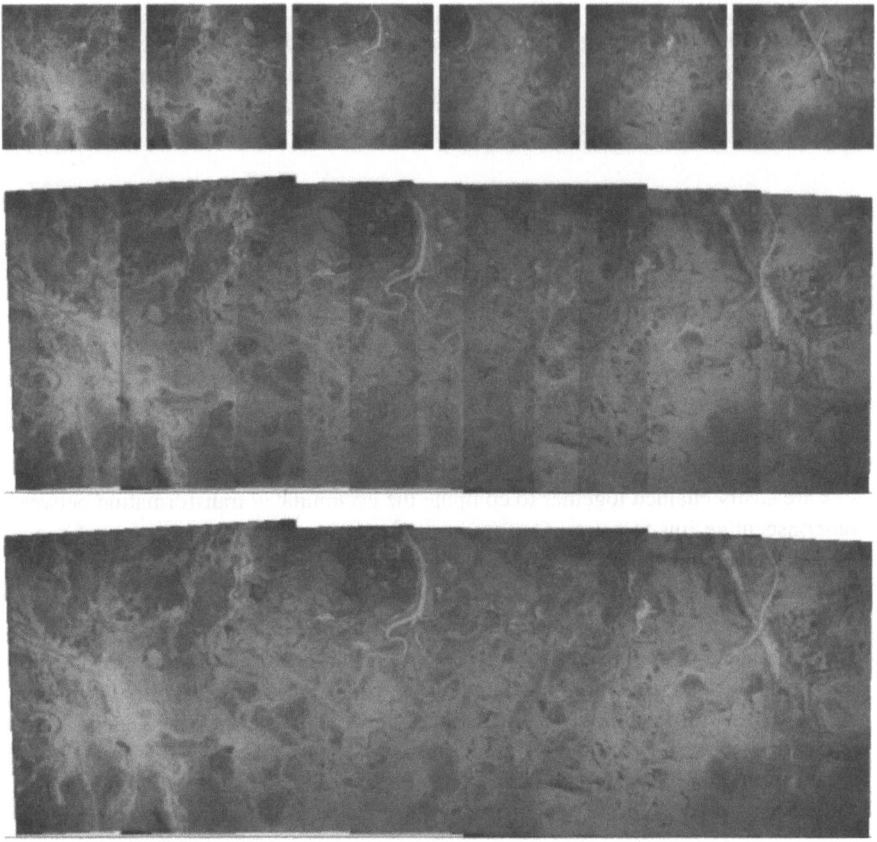

Figure 4.8: (Top) Six aeriel images of the Niger delta region. The images show significant shading toward the periphery, probably due to vignetting effects.(Middle) A mosaic composed from the six images using simple averaging. The shading causes very noticeable seams in the mosaic. (Bottom) Using feathered blending, the seams are ameliorated, but the mosaic still has a "tiger stripes" pattern of brightness variation.

4.3.5 A fast alternative to median filtering

A drawback with the temporal median filtering method of blending is that it is computationally very expensive, requiring a sort operation for every pixel in the rendered image, which can become prohibitively expensive when large numbers of overlapping images are involved. An alternative, which has not been explored in the literature, is to use the *mode* instead of the *median* as a robust estimate of the mean. This has the additional advantage of being able to recover the correct pixel value in cases where the background is occluded by an independently moving object in more than half the frames in which the pixel is visible.

One algorithm which has proved successful blends the images using a robust average which is guided by an estimate of the mode at each pixel. The method for 8-bit, monochrome images is as follows.

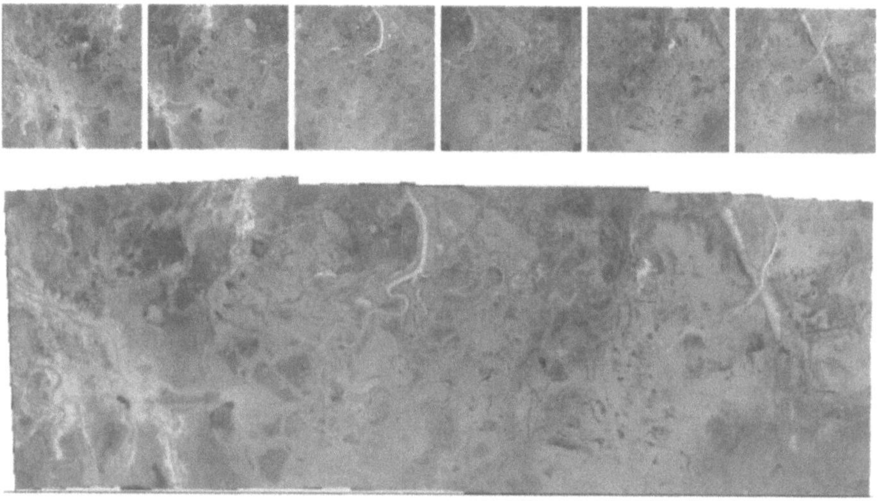

Figure 4.9: (Top) The image sequence shown in figure 4.8 after automatic shading correction. (Bottom) The mosaic composed using the corrected images is both seam-free and of uniform brightness.

1. At each pixel in the rendered image, bucket the values extracted from the input images to form a coarse histogram of overlapping bins. For example, 31 bins, each of width 16, centred on every 8th grey-level.

2. Find the modal bin, and take the average of the values it contains to be the rendered pixel value.

This estimate can be computed very efficiently, even when the number of overlapping images is very large, since the number of histogram bins is fixed, and the mean within each bin requires only a single accumulator per bin. For colour images, the different channels may be processed separately, and recombined to form the final image.

4.4 Simultaneous registration of multiple views

In this section we consider the problem of error accumulation when chaining together homographies over a sequence. This problem can be solved by *simultaneously* registering all of the images into a single, global frame. There are two principal steps necessary to achieve this goal: extension of the two-view ML homography estimator to the *N*-view case; and identification of feature point correspondences over all views, termed the *N-view matching* problem, for which we describe a novel algorithm.

4.4.1 Motivation

The image sequences seen in the previous sections have been simple, single direction camera pans. By constrast, the sequence shown in Figure 4.10 shows a planar scene

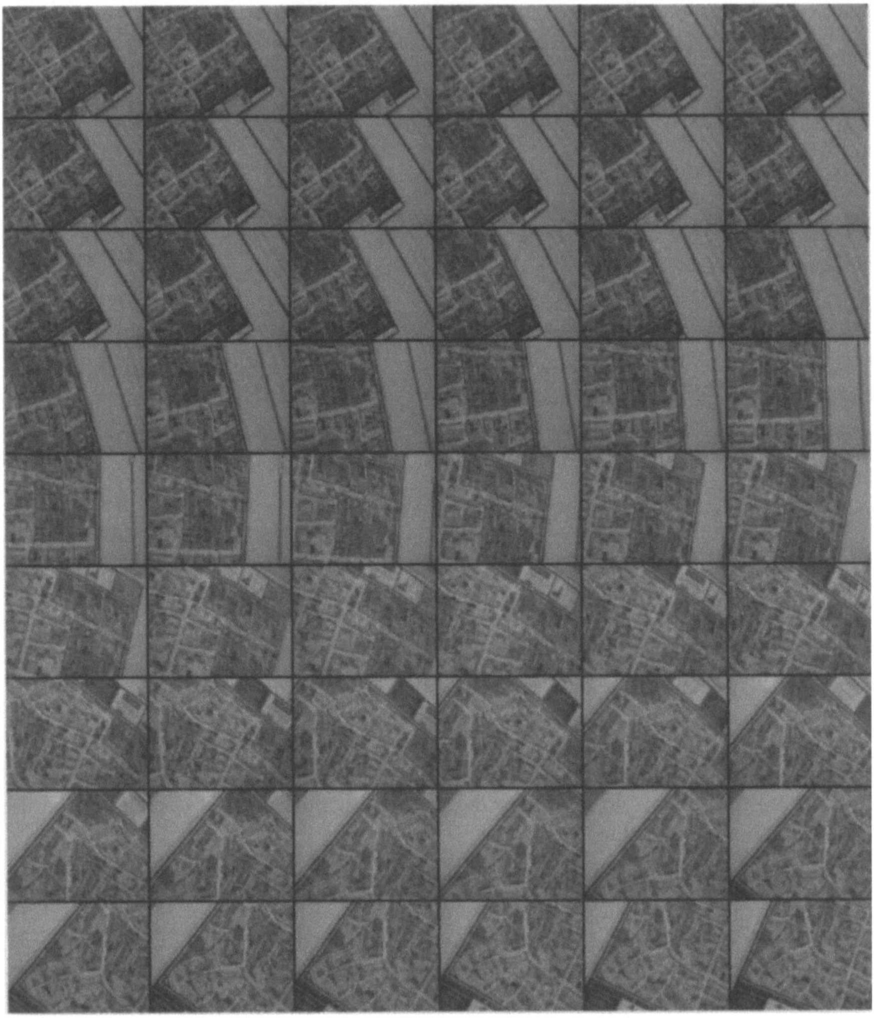

Figure 4.10: The "Oxford Map" sequence of 56 images, showing a planar scene captured using a hand-held digital video camera. The sequence loops around, re-visiting some parts of the scene more than once at different points in time.

captured using a hand-held digital video camera. The key feature of this sequence is that it loops around, re-visiting some parts of the scene more than once at different points in time: a large portion of the first frame is also visible in the last.

Figure 4.11 shows an image rendered from the mosaic representation of the sequence, using the nearest image-centre blending method. The close-up view clearly shows a mismatch between the first and last frames. Figure 4.12 shows the same mosaic, rendered using median filtering, which in this case can do little to ameliorate the misregistration. The problem is that, despite the fact that the two frames are spatially adjacent, with a high degree of overlap, the homographies mapping them into

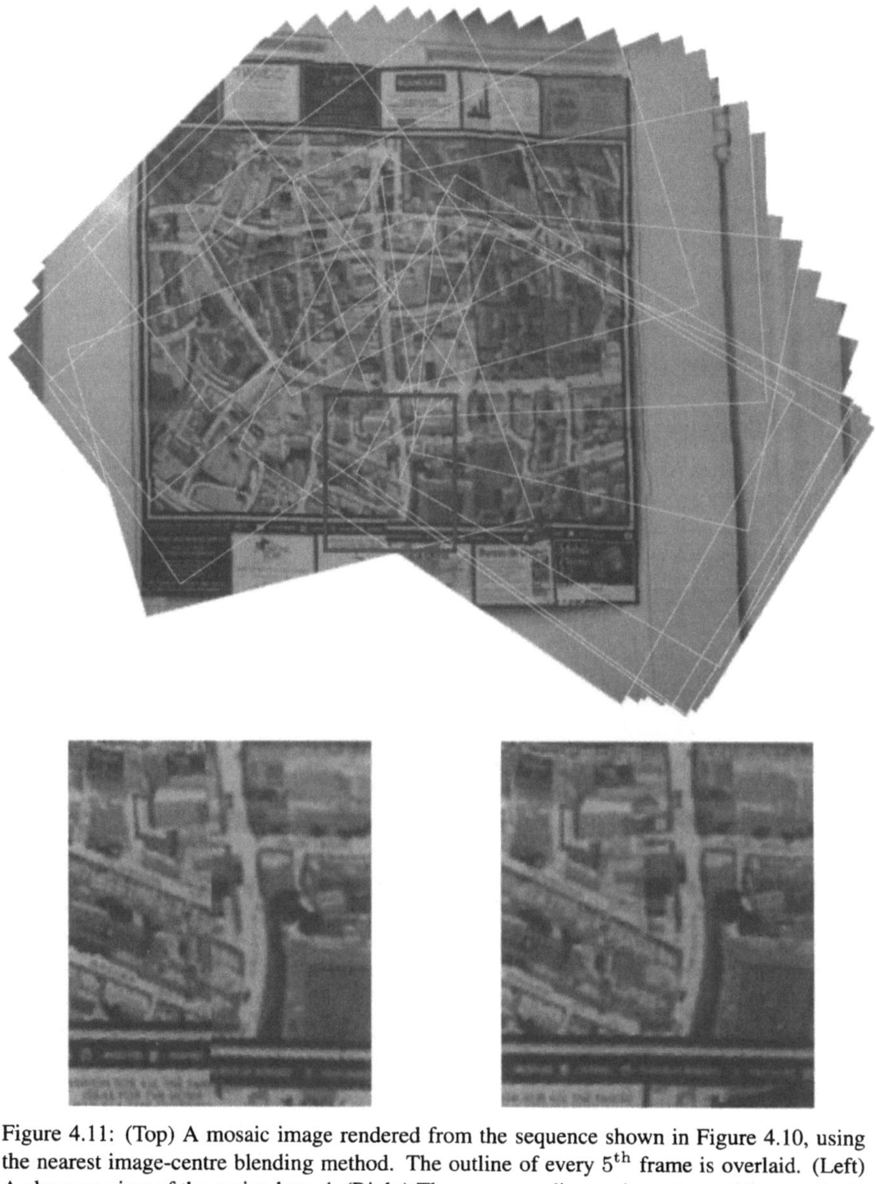

Figure 4.11: (Top) A mosaic image rendered from the sequence shown in Figure 4.10, using the nearest image-centre blending method. The outline of every 5th frame is overlaid. (Left) A close-up view of the region boxed. (Right) The corresponding region extracted from a single frame in the sequence. Comparison of the close-up views reveals a clear mismatch between the first and last frames in the sequence. This is caused by the accumulation of registration error over the sequence.

the mosaic image were computed by concatenating the homographies between the intervening consecutive frames, as described in Section 4.2. This permits registration errors to accumulate rapidly, resulting in a kind of "dead-reckoning" error.

Figure 4.12: The mosaic shown in Figure 4.11, but rendered using median filtering. The result is blurred in areas where registration between temporally distant images is poor.

Various authors have suggested methods for ameliorating the effect of accumulated registration error. Mann and Picard [98] suggest breaking the sequence into smaller sub-sets of frames which are used to create sub-mosaics. The sub-mosaics are then registered and combined to form the final mosaic. Davis [45] solves a linear system, derived from a redundant set of pairwise homographies, so as to minimize an algebraic residual defined over the actual $\mathtt{H}$ matrix elements. Sawhney *et al.* [126, 128] propose a scheme in which the mosaic image is updated one frame at a time, and each additional frame is registered with and blended into the current mosaic image.

Although these are all very practical methods, they are sub-optimal. The optimal solution, as has been known to photogrammetrists for many years [140], is to use *block bundle-adjustment*. This is the method that we adopt. Hartley [75] describes a feature-based bundle-adjustment scheme for the estimation of homographies. Sawhney *et al.* [127] describes an analogous intensity-correlation based method. The latter is rather less elegant than the former, requiring a multi-scale approach combined with a progression of motion-model complexities.

Bundle-adjustment is a direct extension of the two-view maximum-likelihood registration framework to the multiple-view case. In the following sections, we first describe the generalization of the two-view framework to the N-view case, and derive the corresponding N-view ML estimator. We then describe a novel and efficient solution to the problem of matching point features across multiple views, which is a pre-requisite to bundle adjustment.

4.4.2 Extending the two-view framework to N-views

The N-view homography estimation problem can be described as follows. We have N overlapping views of a 3D scene. Points in the views are related to points in the scene by homographies. The objective is to register every view with some global frame containing the whole scene. The choice of global frame is unimportant, but for convenience we shall assume that it is aligned with one of the N images, called the *reference image*.

Now that we are dealing with more than two views, we shall adopt the following notation :

- $\mathbf{x}_i^n$ is the ith interest point observed in the nth view.
- $\mathtt{H}^n$ is the homography which transforms points in the global frame into the nth view.

Every interest point $\mathbf{x}_i^n$ is a noisy observation of a projected pre-image point $\mathbf{X}_j$. Since the views overlap, this pre-image point $\mathbf{X}_j$ will typically also project into several other views, generating other interest points $\{\mathbf{x}\}$. The set $\mathcal{M}_j = \{\mathbf{x}_i^n | \mathbf{x}_i^n \approx \mathtt{H}^n \mathbf{X}_j\}$ is the set of interest points in all views which correspond to the same underlying pre-image point $\mathbf{X}_j$. This is termed an *N-view match*. The situation is illustrated in Figure 4.13, which shows four views of a planar scene. The same pre-image point – marked with a cross – is observed in each view. These four interest points constitute an N-view match. It is important to note that, in general, the number of points in an N-view match $\mathcal{M}_j$ will be less than the number of views, since the pre-image $\mathbf{X}_j$ is only observed in a sub-set of the views. Of course, there will typically be several hundred or several thousand N-view matches in a given set of overlapping images.

Clearly, the concept of an N-view match is a direct extension of the two-view matches already seen, in which certain pre-image points generate corresponding interest points in both images. The derivation of the extended maximum likelihood estimator over all views and all N-view matches also follows directly, as we shall now see.

The negative log-likelihood of the N-view matches $\mathcal{M}_j$ given the N scene-to-image homographies $\mathtt{H}^n$, and M pre-image points $\mathbf{X}_j$ is

$$\begin{aligned} L &= -\sum_{j=1}^{M} \sum_{\mathbf{x}_i^n \in \mathcal{M}_j} \log \Pr(\mathbf{x}_i^n | \mathtt{H}^n) \\ &= \sum_{j=1}^{M} \sum_{\mathbf{x}_i^n \in \mathcal{M}_j} d^2(\mathbf{x}_i^n, \mathtt{H}^n \mathbf{X}_j) \end{aligned} \tag{4.3}$$

where $d^2(\mathbf{x}, \mathbf{x}')$ is the squared (inhomogeneous) Euclidean distance between homogeneous points $\mathbf{x}$ and $\mathbf{x}'$.

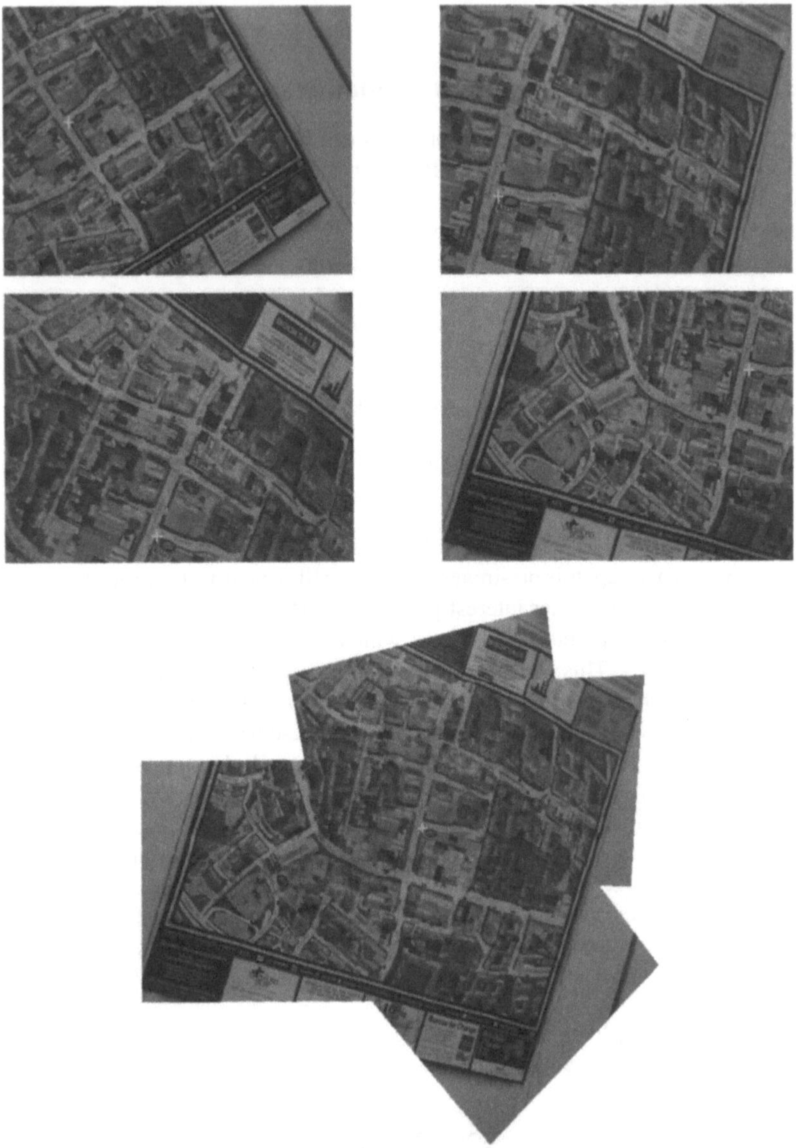

Figure 4.13: (Top) Four views of a planar scene. A particular interest point, marked with a cross, is observed in all four views. (Bottom) When the views are transformed into a global frame (aligned with the 4th image) it is clear that the four interest points are observations of the same underlying pre-image point and that they therefore constitute an *N-view match*.

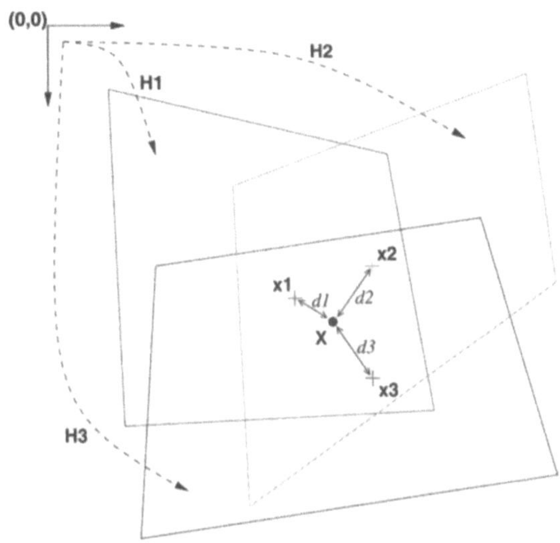

Figure 4.14: The pre-image point $\mathbf{X}$ generates interest points $\mathbf{x}_1$, $\mathbf{x}_2$ and $\mathbf{x}_3$ in three different views. The N-view ML estimator minimizes the distances d_1, d_2 and d_3 with respect to the homographies $\mathtt{H}_1$, $\mathtt{H}_2$ and $\mathtt{H}_3$ and the point $\mathbf{X}$.

The ML estimate of the homographies $\mathtt{H}^n$ is obtained by minimizing L over *both* the parameters of the homographies *and* the positions of the pre-image points $\mathbf{X}_j$. The schematic example Figure 4.14 shows a pre-image point $\mathbf{X}$ which generates interest points $\mathbf{x}_1$, $\mathbf{x}_2$ and $\mathbf{x}_3$ in three different views. The distances d_1, d_2 and d_3 are to be minimized with respect to the homographies $\mathtt{H}_1$, $\mathtt{H}_2$ and $\mathtt{H}_3$ and the point $\mathbf{X}$.

Efficient optimization Unlike the two-view case, there exists no approximation to the N-view log-likelihood which would allow us to avoid explicit parameterization of the pre-image points $\mathbf{X}_j$. The total number of parameters to be estimate is therefore

$$\begin{aligned} \text{number of parameters} &= 8 \times \text{number of views} + 2 \times \text{number of N-view matches} \\ &= 8N + 2M \end{aligned} \tag{4.4}$$

Every interest point $\mathbf{x}_i^n$ is a member of either 0 or 1 N-view match. Each point that *is* a member of some $\mathcal{M}_j$ contributes error residuals $(x_i^n - X_j^n, y_i^n - Y_j^n)$. As in the two-view case, this system of equations constitutes a non-linear least-squares problem, for which Gauss–Newton style iterative algorithms, such as the ubiquitous Levenberg–Marquardt, provide an efficient means of optimization. However, at first glance, the size of the problem appears to prohibit the use of these methods: a typical example may feature 100 views, 1000 N-view matches, and 10 000 interest points, requiring 2800 parameters to be optimized over 20 000 residuals. Fortunately, on closer inspection, we note that certain groups of error residuals are independent of certain groups of parameters: the residuals due to interest points in a particular view

n depend only on the parameters of a single homography H^n, and on the sub-set of *N*-view matches which contain points in view n. This means that, given any putative set of parameters $\{\mathtt{H}, \mathbf{X}\}$, the Jacobian matrix of the geometric residuals with respect to the parameters is block-sparse with a predictable sparsity structure.

An implementation of the Levenberg–Marquardt algorithm which takes advantage of this block-structure to allow efficient computation is described in the *Manual of Photogrammetry* [140] and also by Hartley [74, 75] and Triggs *et al.* [161]. However, the bundle-adjustment in this book was carried out using an implementation based on more modern approaches to large-scale non-linear optimization [24, 102]. The implementation is detailed in Appendix A.

4.4.3 A novel algorithm for feature-matching over N-views

We now address the problem of exactly how to go about finding the *N*-view matches, $\mathcal{M}_j$. Initially, we only have matches computed between consecutive pairs of images in the sequence. To initiate the *N*-view matching process we can simply chain together corresponding feature points in consecutive views to form tracks. For instance, if point i in image 1 is matched with point j in image 2, which is in turn matched with point k in image 3, and so on, then these feature points are all observations of the same pre-image point and hence form an *N*-view match.

Unfortunately, feature points have a tendency to "drop-out" occasionally, causing what should be a single long track to become fragmented. Also, a particular pre-image point may be visible for short periods at several points in the sequence, but outside the field of view in the intervening periods, again causing several distinct *N*-view matches to be detected when there should only be one. Clearly, it is unrealistic to expect simple feature tracking to provide any *N*-view matches connecting the first and last frames in our example sequence. An algorithm is required for merging the disjointed tracks. To facilitate this, we introduce the concept of a *view graph.*

The view graph The "view graph" is an undirected graph with images at the nodes, and homographies and two-view point-correspondence sets on the edges. Having performed registration between consecutive images in a sequence, the initial topology of the view graph is linear, with edges between temporally adjacent images. Homographies between temporally separated images may be computed by following a path through the graph from node i to node j, chaining together the homographies on the edges. Similarly, the two-view matches may be chained together to form tracks. Performing explicit two-view registration between any pair of images allows a new edge to be inserted into the graph, admitting the possibility of merging tracks which correspond to the same pre-image point, and which pass through both images.

The view-graph concept allows us to explicitly define what we mean by a *globally consistent* set of homographies :

> *A globally consistent set of homographies is such that an aggregate homography computed by concatenation around any cycle in the view graph is equal to the identity.*

A set of homographies computed between overlapping frames using the two-view algorithm does not possess this property due to the error accumulation problem. In contrast, the *N*-view ML estimator guarantees global consistency.

Explicit registration of every image with every other image, so as to form a fully connected graph, will allow every stray track to be merged with its siblings, but clearly this is infeasible for large numbers of images. An efficient *N*-view matching algorithm will add the minimum number of additional edges to the view graph such that all the fragmented tracks are correctly merged into unique *N*-view matches. This is the aim of the novel algorithm, which is now described.

The concept of a graph structure, whose topology represents the explicit pairwise registrations performed within a set of images, was also introduced independently by Sawhney *et al.* [126]. However, whereas we are interested in propagating match-tracks across spatially adjacent frames, and hence mostly interested in the point correspondences attached to the graph edges, they are interested in the actual homographies attached to the graph edges and in the choice of shortest-path routes along which to concatenate homographies in an effort to minimize error accumulation. They use an intensity-based registration method, and hence have no concept of *N*-view match tracks. Consequently, their heuristic rules for deciding which images pairs to explicitly match are rather different to the ones described below.

The key to our algorithm is the selection of a good criterion for deciding how beneficial it may be to explicitly register any given pair of images. Our criterion is based on the following observation :

> *Image pairs which have a high degree of overlap are likely to yield many matched feature points. Therefore any such pair of images which additionally share few N-view match tracks are good candidates for explicit matching.*

The important factors are therefore the area of overlap between a pair of images, and the number of *N*-view matches which pass through both of them. It is clearly pointless to try to match images with very little overlap. It is also of little benefit to match images which already have many shared match tracks.

The area of overlap between a pair of images (i, j) is computed as follows. We first compute the homography $\mathtt{H}_{(j \rightarrow i)}$ by following the current shortest path between nodes i and j in the view graph. The boundary of image j is then transformed into and intersected with the boundary of image i. We define the relative area of overlap as

$$a_{ij} = \frac{\text{area of intersection}}{\max\left(\text{area of image } i, \text{area of image } j\right)} \tag{4.5}$$

a is a number between 0 and 1, where 1 indicates perfect overlap. A simple function which captures the heuristic rules given above is

$$C_{ij} = a_{ij} \exp(-M_{ij}/k) \tag{4.6}$$

where M_{ij} is the number of *N*-view matches which contain feature points from both image i and image j, and k is a constant which controls the relative importance of intersection area and number of mutual matches ($k = 100$ is a ball-park value). This function has a high value for image pairs which are considered to be good candidates for explicit matching. The N-view matching is outlined in Table 4.1.

Algorithm
1. Initialize the N-view matches by linking together pairwise matches between consecutive frames.
2. Using the current view graph, rank every possible pair of images according to the criterion given in Equation (4.6).
3. Perform explicit registration between the highest ranking pair, using the algorithm of Section 3.4.
4. Incorporate the new edge into the view graph, and merge N-view matches where indicated by the newly found pairwise matches.
5. Repeat steps 2, 3 and 4 until the total number of N-view matches stabilizes.

Table 4.1: The main steps in the N-view matching algorithm.

When merging tracks, it is important to ensure that the tracks being merged are consistent. If each contains a different feature point in the *same* view then clearly they cannot be merged, indicating that a mismatch has occurred somewhere. In this event, both tracks are discarded. However, since the outlier rejection capability of the two-view matching algorithm is very powerful, such mismatches are extremely rare.

Plate 10(a,b) show the view graph before and after application of the N-view matching algorithm. Plate 10(c,d) represent the number of feature matches between every pair of frames as a colour-coded matrix. Table 4.2 shows the distribution of track lengths before and after N-view matching. The number of long tracks has increased dramatically and there are now tracks linking the first and last frames in the sequences. 45 edges have been added to the graph.

4.4.4 Results

Having found a good set of N-view matches, we are in a position to compute a *globally consistent* set of homographies using the N-view ML estimator. Being a nonlinear method, the estimator requires initial estimates of both the homographies $\{\mathtt{H}\}$ and the pre-image points $\{\mathbf{X}\}$ (of which there will be one per N-view match). The homographies are initialized by concatenating along shortest-paths in the view graph that is generated during the N-view matching process. The pre-image points are initialized by selecting any one of the feature points from each N-view match set, and transforming it into the reference frame.

Figure 4.15 shows the "Oxford Map" mosaic computed using both the original homographies (computed by concatenation), and the globally consistent homographies

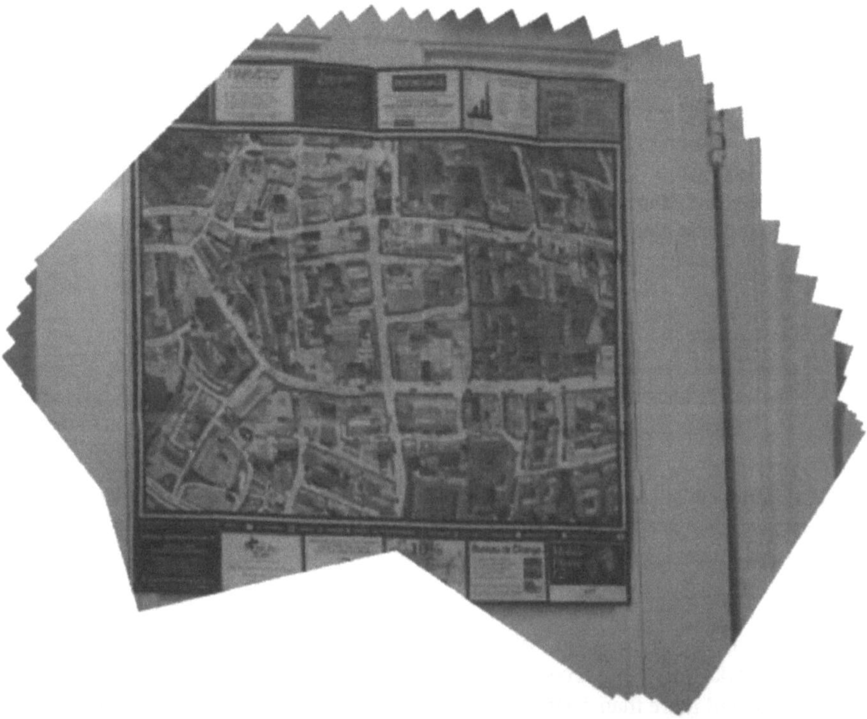

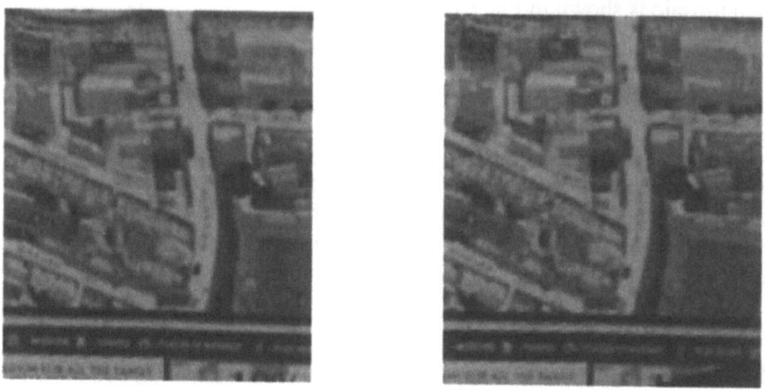

Figure 4.15: (Top) The mosaic shown in figure 4.11 after refinement of the homographies by bundle-adjustment. (Left) A close-up view of the boxed region shown in Figure 4.11. (Right) The corresponding region extracted from a single frame in the sequence. Comparison of the close-up views shows that the mismatch visible in the original mosaic has been eliminated.

obtained using bundle-adjustment. The close-up views clearly show that the mismatch between the first and last frames has been eliminated.

Figure 4.16 shows a sequence of 94 images of the University Museum, captured using a MiniDV camera. Plate 11 shows the view graphs for this sequence before and after N-view matching. Table 4.3 shows the distribution of N-view match track lengths

Distribution of track lengths	> 10 frames	> 20 frames	> 30 frames
Before N-view matching	561	74	9
After N-view matching	804	175	18

Table 4.2: **Oxford Map sequence** The distribution of N-view match track lengths before and after the automated N-view matching algorithm. N-view matching allows short tracks to be spliced together across the sequence. The number of tracks longer than 10 frames is increased by 43%.

Distribution of track lengths	> 10 frames	> 20 frames	> 30 frames	> 40 frames
Before N-view matching	686	132	39	0
After N-view matching	1383	475	189	51

Table 4.3: **University Museum sequence** The distribution of N-view match track lengths before and after the automated N-view matching algorithm. The number of tracks longer than 10 frames is more than doubled, and over 50 tracks are longer than 40 frames.

before and after matching. In this example, the heuristic for choosing which images to match works very well. The number of match tracks covering more than 10 frames is doubled, and more than 50 tracks are longer than 40 frames. The completed, bundle adjusted mosaic is shown in Plate 12.

4.5 Automating the choice of reprojection frame

In this section we consider the problem of rendering "bow-tie" style, planar manifold mosaics from rotating camera sequences. We describe an automatic method for choosing the rendering transformation which minimizes the projective distortion of the peripheral images, producing a well-balanced and aesthetically pleasing mosaic.

4.5.1 Motivation

As explained in Section 4.2, the task of the rendering transformation is to map points in the rendered image into points in the global reference frame, and hence into the individual images. The global reference frame is arbitrary, and for convenience, is generally aligned with one of the input images, called the *reference image*. In the case of a planar reprojection manifold, the simplest choice of rendering transformation is a similarity transformation, the purpose of which is to shift the origin of the global reference frame to the origin of the rendered image, and to scale the rendered image to the desired dimensions. The rendering transformation from the rendered image into image k is then

$$\mathtt{T}_k = \mathtt{H}_{(\mathrm{ref}\rightarrow k)}\mathtt{S} \tag{4.7}$$

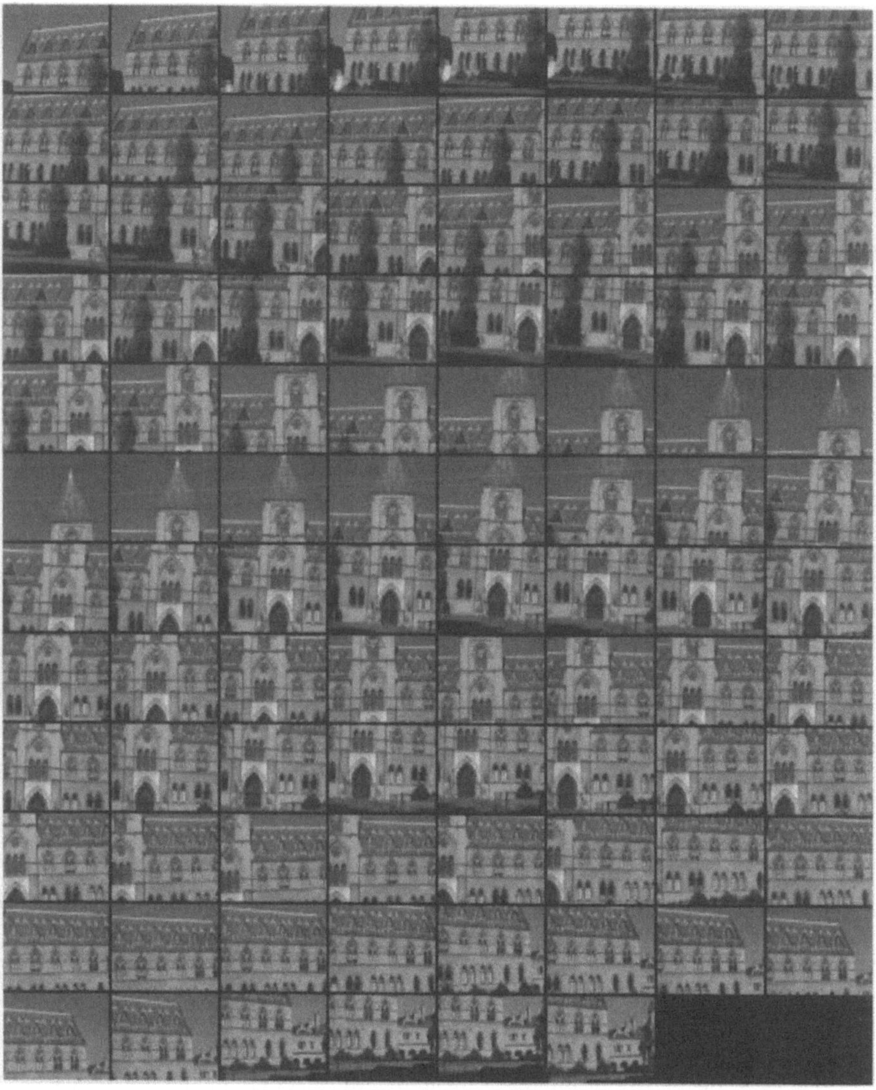

Figure 4.16: The University Museum sequence. There are 94 frames in total, captured using a mini-DV camera.

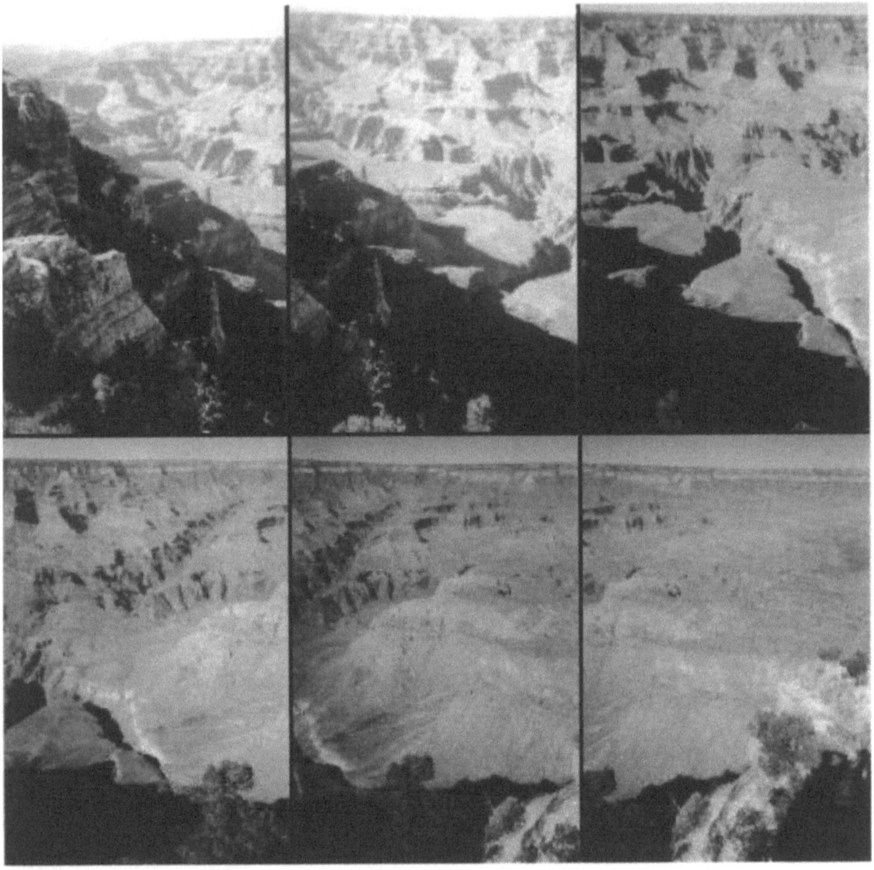

Figure 4.17: A sequence of six images of the Grand Canyon, taken with a hand-held 35 mm camera and scanned using a flat-bed scanner.

where the similarity transform S combines the camera calibration matrices of the real reference camera and the virtual camera :

$$S = K_{\text{ref}} K_{\text{virtual}}^{-1} \qquad (4.8)$$

In this case, the choice of reference image is critical to producing a well-balanced mosaic, and a good choice is generally an image near the middle of the sequence.

The method is demonstrated in Figure 4.18, which shows two mosaics made from the sequence of six images shown in Figure 4.17. Given the small number of images, our choice of reference frame is very limited. The first mosaic is composed using the third image as the reference frame, and the second mosaic uses the fourth image. The outlines of the images are superimposed on each mosaic. Clearly, neither choice provides a satisfactory reference frame in which the projective distortion is balanced.

Figure 4.18: Mosaics composed from the six images shown in Figure 4.17. (Left) The third image is used as the reference frame. (Right) The fourth image is used as the reference frame. The third and fourth images are the closest to the middle of the sequence, yet neither provides a satisfactory reference frame in which the projective distortion is balanced.

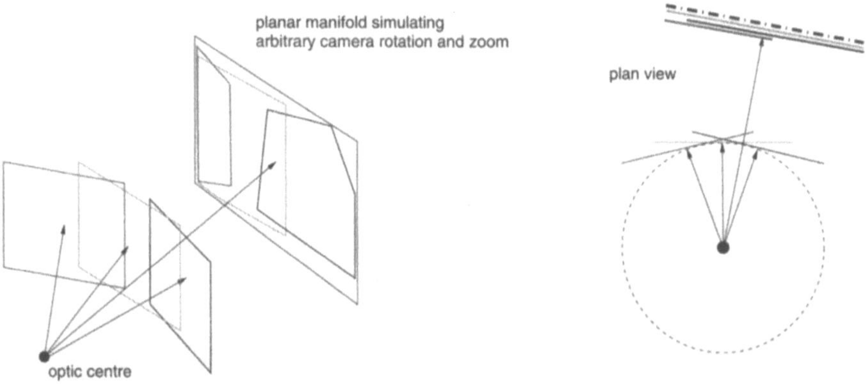

Figure 4.19: Schematic illustration of using planar manifold projection to simulate an arbitrary camera rotation and zoom.

4.5.2 Synthetic camera rotations

In choosing a reference image, we are implicitly setting the optic axis of our virtual camera to be the same as that of the real camera which captured the reference image. It is perfectly possible however, to create a virtual camera, concentric with the real camera, but with any chosen gaze direction and focal length. This idea is used to great effect in systems such as the Quicktime VR software [35], which gives the user the sensation of standing at a single spot (referred to as a node) from where he is free to look in any direction and to zoom in or out. In such systems, the environment is typically represented by a cylindrical or spherical mosaic, and rendering is by reprojection of a small portion of the view sphere onto a planar manifold. The scheme is illustrated schematically in Figure 4.19.

Referring to equation (4.2), the model of a rotating projective camera, we can write the homography relating any pair of images (i, j) as a conjugate rotation:

$$\mathtt{H}_{(i \to j)} = \mathtt{K}_j \mathtt{R}_{(i \to j)} \mathtt{K}_i^{-1} \tag{4.9}$$

Given the calibration matrix for each image $\mathtt{K}_n$, along with the desired internal parameters $\mathtt{K}_{\text{virtual}}$ and rotation $\mathtt{R}_{\text{virtual} \to \text{ref}}$ of the required virtual camera, the appropriate

Figure 4.20: The same mosaic as shown in Figure 4.18 rendered using a variety of synthetic camera rotations, about the x-axis (top), y-axis (middle) and z-axis(bottom).

rendering transformation, which maps points in the mosaic image to points in image k, may be computed using equation (4.9) :

$$\begin{aligned} \mathtt{T}_k &= \left(\mathtt{K}_k \mathtt{R}_{\text{virtual}\to k} \mathtt{K}_{\text{virtual}}^{-1}\right) \\ &= \mathtt{H}_{(\text{ref}\to k)} \left(\mathtt{K}_{\text{ref}} \mathtt{R}_{\text{virtual}\to\text{ref}} \mathtt{K}_{\text{virtual}}^{-1}\right) \end{aligned} \tag{4.10}$$

Unlike the QuickTime VR situation, we are interested in reprojecting *all* of the frames in a sequence onto a planar manifold (in cases where this is possible.) Figure 4.20 shows the Grand Canyon mosaic rendered under various different virtual camera rotations. As the rotation varies, so does the projective distortion of the reprojected images. Our novel algorithm automatically chooses a rotation which minimizes this projective distortion, resulting in a well-balanced rendering.

There are various ways in which we might quantify the level of distortion, but the one that has been found particularly effective is to compute the ratio of the maximum and minimum reprojected image area :

$$\begin{aligned} A_k &= \text{area}(\ \mathtt{T}_k^{-1} \times \text{image boundary}\ ) \\ C &= \frac{\max\{A_k\}}{\min\{A_k\}} \end{aligned} \tag{4.11}$$

A well-balanced mosaic minimizes the cost function C (the optimal value being 1). As can be seen in Figure 4.20, rotation about the z-axis (cyclotorsion) has no effect on the level of projective distortion, so optimization is only performed over the x- and y- axis rotations (yaw and pitch). Correct application of cyclotorsion requires

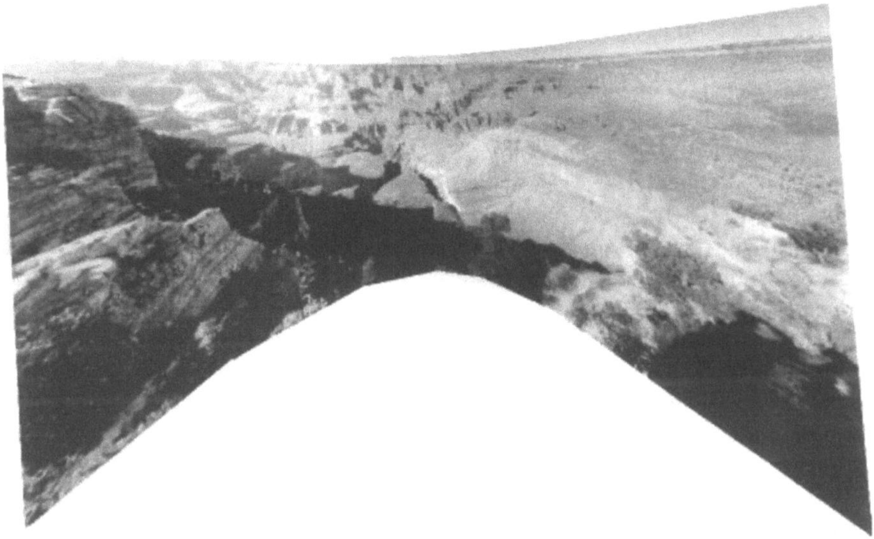

Figure 4.21: The mosaic sequence of Figure 4.18, but with the rendering transformation computed by the automatic balancing algorithm.

understanding of the correct orientation of objects in the scene, and hence it is best left to the user to apply as a post-processing step.

The automatic balancing algorithm is as follows.

- **Initialization** Initialize the yaw and pitch angles to zero. From all the images in the sequence, choose a reference image which gives the minimum value of the cost function of equation (4.11).
- **Calibration** Estimate the focal length of the real reference camera using de Agapito's linear method.
- **Optimization** Further minimize equation (4.11) with respect to the yaw and pitch angles using gradient descent. Iterate until the minimum is reached.
- **Post-processing** Find the bounding box of the rotated mosaic, and compute the similarity transformation S which maps the bounding-box onto the desired rendered image (dimensions specified by user). Render the mosaic using the chosen blending method.

The result of applying the algorithm to the image sequence shown in Figure 4.18 is shown in Figure 4.21. The mosaic is now well balanced and pleasing to the eye.

4.6 Applications of image mosaicing

Irani *et al.* [82] provide a good review of 2D mosaicing applications. Some of these rely on the idea of computing *significant residuals* between input frames and the mosaic. Such residual motions are usually the result of off-plane structure. These regions are segmented out using a temporal median filter. One application of this is off-line video compression using a *static mosaic*. The static mosaic consists of a single mosaic constructed by a temporal median and representing the dominant background. In addition to this, the significant residuals for every frame are also extracted and, along with the mosaic image, can be compressed using standard techniques to form a very compact representation of the sequence — a point capitalized on in MPEG4.

Further examples of off-line applications include event synopsis [81] and video-editing [9, 79]. Event synopsis involves constructing a static mosaic and significant residuals and then overlaying the residuals for a number of frames onto the mosaic simultaneously. This creates a synopsis of foreground events over an extended time period. This could be used to create "strobe-effect" tracks of the players and ball at crucial moments in a football game. Digital video composition can be greatly simplified by mosaicing because it effectively allows the user to operate on many frames at once, removing the tedious task of (say) repositioning a matte mask in a whole sequence of frames.

On-line applications include low bit-rate transmission of video [83] and an artificial "steady-cam". These depend on building a *dynamic mosaic*. Such a mosaic is constructed by continually updating with the current incoming image by calculating the significant residuals between the current mosaic and the current image and using them to immediately modify the mosaic. The significant residuals are generally much smaller than for the static mosaic and hence this method is well suited to low bit-rate transmission, requiring only the registration parameters and compressed residuals to be transmitted for each frame.

The dynamic mosaic is also well suited to "steady-cam" applications since most of the high-frequency jitter in an unstabilized camera is due to small rotations. By writing each frame into a dynamic mosaic it is possible to show a steady view and to synthesize smooth camera rotations by appropriately warping the mosaic.

4.7 Mosaicing non-planar surfaces

The essence of chapter 3 is that, for particular camera motions or types of scene, images may be registered by a global mapping with a small number of parameters. Once the parameters of the mapping are determined, the mosaicing problem is reduced to image rendering. As mentioned in chapter 3, parameterized mappings can be derived for surfaces other than planes, such as quadrics or surfaces of revolution, without any requirement for camera calibration, thereby allowing mosaics to be composed from images of such surfaces.

4.8 Mosaicing "user's guide"

The following paragraphs summarize the different imaging situations to which each mosaicing and blending method previously discussed is best suited.

Planar mosaicing Reprojection onto a planar manifold emulates the view from a single, perspective camera with a very large imaging sensor and wide field of view. It requires that the camera motion be a true nodal pan, or alternatively that the scene be purely planar or very distant. In practice, this projection is only really useful when the field of view swept by the camera is < 90 degrees; any greater and the result is a mosaic which exhibits unacceptably severe projective distortion of the source images. It does however have the advantages of preserving straight lines and perspective projection.

Cylindrical mosaicing Reprojection onto a cylindrical or conical manifold emulates a "panoramic" camera in which the film is wrapped in a cylinder about the optic centre. The camera motion is required to be a true nodal pan, and estimates of the camera's internal parameters are also required in order for this method to work correctly. It has the advantage over planar mosaicing that it may be applied to sequences with up to 360 degrees camera rotation. However, it has the drawback that straight lines in the scene map to curves in the mosaic, and the familiar impression of perspective is lost.

Peleg's "manifold projection" This method may be used to generate mosaics from a camera undergoing general motion, provided that a method exists to estimate the dominant motion between consecutive frames. It is hence quite forgiving to deviations from the nodal pan motion model. The mosaics so produced are similar to what might be generated by a 1D "push-broom" camera. Its robustness makes it a useful method for consumer "desktop video" applications such as *VideoBrush*. It may be used with camera rotations up to 360 degrees, as well as camera translations. It has the drawbacks that straight lines map to curves, and the impression of perspective is lost. Furthermore, unlike planar or cylindrical mosaics, the "manifold projection" does not represent the view from any sort of real, single camera. It is therefore of little use in applications such as node-based virtual reality (e.g. Quicktime VR).

4.9 Summary

In this chapter the basic principles of forming a mosaic representation of a scene, and of rendering an image from that representation have been explained. The two-view maximum likelihood homography estimator described in Chapter 3 has been extended to perform simultaneous estimation of homographies between multiple overlapping views. This N-view ML estimator has been shown to find a globally consistent set of homographies, thereby eliminating the problem of error accumulation when concatenating homographies between temporally distant frames. The photometric registration method described in Chapter 3 has been shown to be effective in eliminating seams and balancing colours in rendered mosaic images. A novel algorithm has been described for automatically choosing the gaze direction of the virtual camera so as to minimize projective distortion when rendering a planar manifold mosaic.

4.9.1 Further examples

Plates 13, 14 and Figure 4.22 show some more examples of mosaics rendered using the algorithms described in this chapter.

Figure 4.22: A complete 360° panoramic mosaic composed from 178 frames and rendered using cylindrical projection.

Chapter 5

Super-resolution: Maximum Likelihood and Related Approaches

5.1 Introduction

In this chapter we investigate the problem of fusing information from several views of a scene in order to reconstruct a novel view of the scene with greater spatial resolution than is available in any of the observed views.

In each view, light emitted from the scene is projected onto the sensor array of the camera, producing an irradiance pattern which is temporally integrated, spatially sampled and quantized to generate an image. In addition, the image is often blurred due to de-focus and/or motion.

We attempt to reverse these degradations in order to estimate a high resolution representation of the intensities in the scene. Under the assumption of Lambertian surfaces and uniform illumination, this is equivalent to estimating surface albedo. These techniques are often termed *super-resolution*.

In Section 5.2 we briefly consider exactly what we mean by the term "resolution". Section 5.3 reviews the basic principle of single image restoration. Section 5.4 discusses in detail the assumptions behind our generative image model, contrasting different implementation approaches, and proposing an accurate and efficient method. Section 5.5 describes a simple experiment which aims to justify the Gaussian blur assumption made in the generative model. Section 5.6 describes the generation of the synthetic image sequences which are used to probe the behaviour of the super-resolution algorithms. Section 5.7 describes the creation of an "average image" from a set of low-resolution images, and presents derivations of its properties which are verified empirically. Section 5.8 briefly reviews the super-resolution method proposed by Rudin *et al.* [122]. Section 5.9 introduces the maximum-likelihood super-resolution estimator, and Section 5.10 discusses its properties. Section 5.11 presents an em-

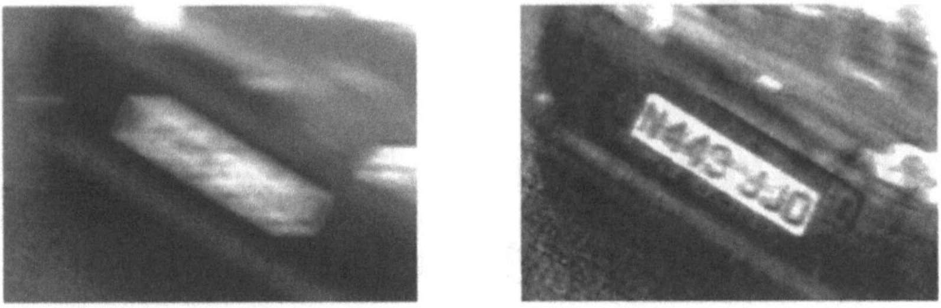

Figure 5.1: (Left) A severely motion blurred car license plate. (Right) After deblurring, the perceived resolution is clearly much improved, although the number of pixels has not changed.

pirical investigation of the behaviour of the ML estimator. Section 5.12 reviews the super-resolution algorithm of Irani and Peleg [84], and analyses its convergence behaviour and relationship to the ML estimator. Finally, Section 5.13 demonstrates the ML estimator applied to some real image sequences.

5.2 What do we mean by "resolution"?

Considering the types of image sequence to which super-resolution techniques are generally applied, we observe that they usually fall into two broad categories:

- **Aliasing is the principal degredation:** sequences in which the optical or motion blur was negligible, but the required texture details are unavailable because the spatial sampling density of the camera was too low.

- **Blur is the principal degredation:** sequences in which the sampling density would be perfectly adequate to capture the required details, but in which the optical or motion blur is severe.

In most super-resolution literature, the term "resolution" is taken to mean the actual number of pixels present in an image. The goal is then to obtain a high-resolution image with a higher pixel density than any of the low-resolution images. A typical "zoom factor" is a doubling of pixel resolution in the x and y directions, producing a high-resolution image with four times the pixel density of any low-resolution image.

This "pixel counting" notion of resolution enhancement is readily applicable to the aliasing dominated cases. In blur dominated cases however, it is generally unnecessary to increase the pixel density. The required texture details are revealed by removing the blur degradation. A motivating example of this idea is shown in Figure 5.1, which shows a severely motion blurred, and consequently unreadable, vehicle registration plate. The deblurred version represents a dramatic improvement in readability, although the number of pixels has not been increased. In such cases, using multiple images helps to reduce the noise sensitivity of the reconstruction process compared with reconstruction from a single image.

For this reason, we shall use a rather more liberal, perceptually motivated definition of resolution enhancement :

> *The super-resolution image should demonstrate an improvement in the perceived detail content compared with that of the low-resolution images. This will typically involve restoration of the high-frequency content, which in turn may require an increase in pixel density.*

Clearly, this definition is rather subjective. However, we would hope that in most cases, the improvement in perceived detail would be clearly visible to any observer.

5.3 Single-image methods

In this section we motivate the argument for using multiple views by looking at an example of restoration from a single image.

Image restoration from single images has generally concentrated on the removal of optical blur and motion blur. A good survey of these methods is given by Gonzalez and Wintz [70], and by Pratt [115]. The blurring process is modelled as convolution of the original image with a blurring operator (also called the point spread function). Hence

$$g_o(x,y) = g_t(x,y) * h_d(x,y) + n(x,y) \quad (* = \text{convolution}) \tag{5.1}$$

where g_o is the observed image, g_t is the true image, h_d is the blur operator, and n is a noise term. Optical blur is modelled by an isotropic, linear, spatially-invariant convolution kernel. Motion blur due to *pure translational motion* is modelled by a non-isotropic kernel. The simplest approach to recovering the original images is to apply a reconstruction operator h_r which reverses the effect of h_d. Hence

$$\begin{aligned} \hat{g}_t(x,y) &= g_o(x,y) * h_r(x,y) \\ &= (g_t(x,y) * h_d(x,y) + n(x,y)) * h_r(x,y) \end{aligned} \tag{5.2}$$

The analysis becomes easier if we turn to the Fourier domain,

$$\hat{\mathcal{G}}_t = (\mathcal{G}_t\mathcal{H}_d + \mathcal{N})\mathcal{H}_r \tag{5.3}$$

The simplest reconstruction operator is the *inverse filter*, $\mathcal{H}_r = \frac{1}{\mathcal{H}_d}$. This leads to

$$\hat{\mathcal{G}}_t = \mathcal{G}_t + \frac{\mathcal{N}}{\mathcal{H}_d} \tag{5.4}$$

The blur operator h_d is a low-pass filter, so $\lim_{\omega\to\infty} \mathcal{H}_d(\omega) = 0$. The inverse filter is therefore a high-pass filter with the opposite characteristic, $\lim_{\omega\to\infty} \mathcal{H}_r(\omega) = \infty$. This means that the inverse filter greatly amplifies the high frequencies of the noise component $\mathcal{N}$, and this noise tends to dominate the reconstructed image. Figure 5.2 shows the effect of applying an inverse filter to a blurred and noisy image.

A more sophisticated reconstruction operator is used in *Wiener filtering*, a technique which utilizes prior knowledge of the spectral densities of the image noise and of the undegraded image. This allows a reconstruction kernel to be derived which minimizes the restoration error E, where

$$E = \|g_t(x,y) - \hat{g}_t(x,y)\|^2 \tag{5.5}$$

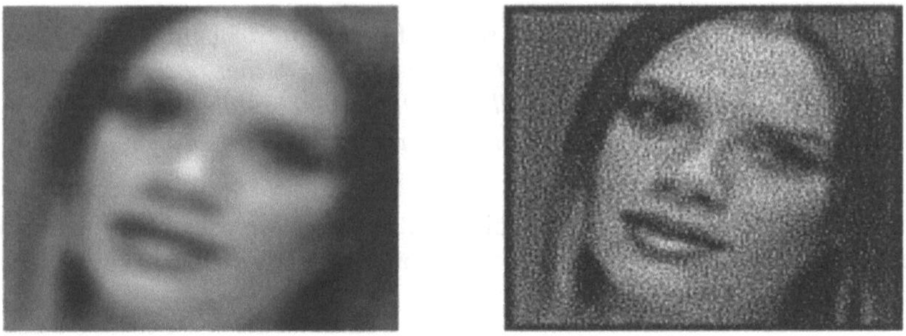

Figure 5.2: A synthetically blurred image and the reconstructed image using the inverse filtering technique. Notice how noise in the blurred image has been amplified in the reconstruction.

The Wiener reconstruction kernel typically has a band-pass rather than a high-pass characteristic, and hence does not suffer from the noise amplification seen in simple inverse filtering. We shall return to the Wiener filter in the context of Bayesian super-resolution methods in Chapter 6.

5.4 The multi-view imaging model

To recover information from several views, we must first model the process by which the individual images are generated. In this section we describe our *generative model*, and then briefly discuss the differences between this model and those adopted elsewhere in the literature. The model used can be factored into several components as follows.

- **The motion model** is planar projective. In other words, the relative motion between the scene and camera induces a projective transformation (homography) between views.

- **The scene** is restricted by our chosen motion model to be either a planar Lambertian surface, or a general static scene viewed by a camera rotating about its optic axis and/or zooming. As shown in Chapter 3, for such scenes the geometric transformation between views is a homography.

- **The illumination** is assumed to be uniform across the surface. It may also vary constantly as the surface and/or camera moves.

- **The camera sensor** is linear device comprising a rectangular array of uniformly spaced CCD elements, each of which integrates the radiant energy falling across its surface for the duration of the open-shutter exposure. If the scene or camera moves during the exposure, motion blur is observed.

- **The camera geometry** is taken to be that of a pinhole camera. That is to say, the scene is projected onto the image plane of the camera by a perspective projection, which is a special case of the general projective transformation.

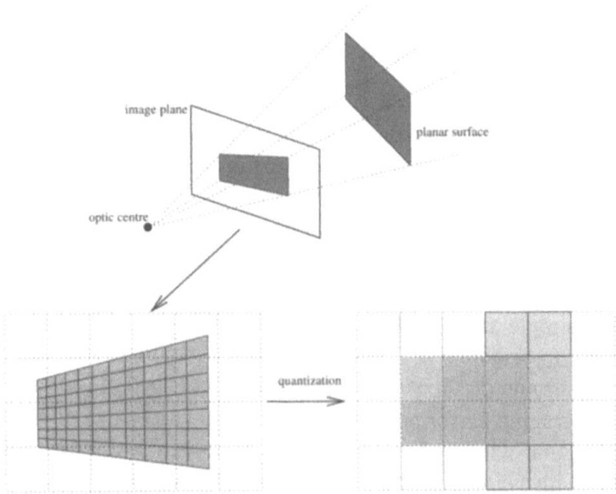

Figure 5.3: (Top) A perspective camera is imaging a planar scene. (Left) The scene is projected onto the CCD array by perspective projection. (Right) The CCD elements are assumed to integrate the radiant energy over their surface, producing a spatially-quantized image of the scene.

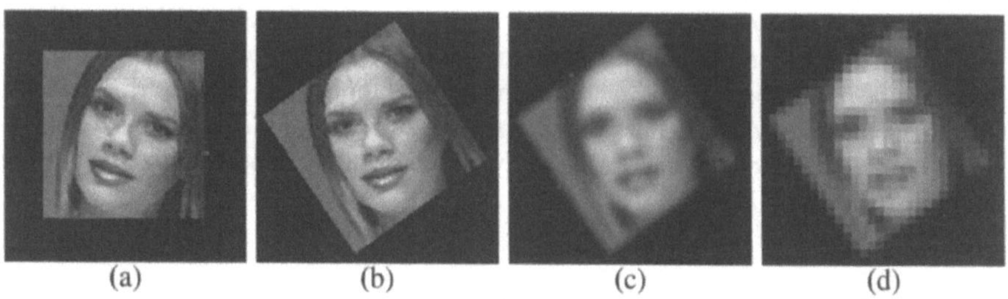

Figure 5.4: The principal steps in the imaging model: the high-resolution planar surface (a), undergoes a geometric viewing transformation (b), followed by optical/motion blurring (c), and finally down-sampling (d).

- **The camera optics** may induce blur due to de-focus. The blur is assumed to occur uniformly across the camera sensor, and is therefore modelled as the convolution of a linear, spatially invariant point-spread function with the irradiance pattern projected onto the sensor plane.

- **The image noise** is assumed to be spatially uncorrelated, isotropic, additive, normally distributed with mean equal to zero, and constant variance.

The situation is shown schematically in Figure 5.3, and the principal steps are illustrated in Figure 5.4.

The image-to-image transformations are estimated *a priori* using the geometric registration techniques described in Chapter 3.

5.4.1 A note on the assumptions made in the model

In addition to the rotating camera and planar scene cases, in which the planar motion model is exact, the planar homography is also a very good approximation of the view-to-view transformation under imaging conditions in which parallax is negligible, i.e. when the surface relief is very small by comparison with the distance to the camera. An example of this is in high aerial or satellite views of the Earth.

The Lambertian model, together with the assumption that illumination varies uniformly across the scene, means that the global intensity mapping between views can be modelled as a simple affine scaling (a multiplicative term akin to contrast, and an additive term akin to brightness). Estimates of these parameters are also obtained *a priori* using the photometric registration techniques described in Chapter 4.

The *spatial* averaging performed by each CCD cell can be modelled as a convolution with a rectangular window [28]. The effect of *optical* blur is typically modelled by a convolution with a circular, "top hat" function [8]. We model the combined effect of the two degredations as convolution with an isotropic Gaussian. Further justification for this model is given in Section 5.5. Note that we do not attempt to model motion blur.

The assumption of a linear mapping between scene irradiance and pixel value is not perfectly true in practice because of the dynamic range compression often performed by some component of the video capture system. Details of this non-linear mapping are usually not available in desktop video capture equipment, although methods have been proposed to estimate it using calibration images [117, 165]. However, such mappings are typically fairly linear for a large portion of the mid-intensity range, a fact observed by Robertson *et al.*in their work on dynamic range improvement using multiple exposures [119].

5.4.2 Discretization of the imaging model

So far, our discussion of the imaging model has assumed that the scene intensities are a continuous two-dimensional function. In order to make the model useful in an optimization context we choose a suitable discrete representation of the scene intensities, and discretize the imaging model accordingly.

It is assumed that the set of observed low-resolution images were produced by a single high-resolution image under the following generative model

$$g_n(x,y) = \alpha_n \; s{\downarrow}\left(h(u,v) * \bar{f}(\mathcal{T}_n(x,y))\right) + \beta_n + \eta(x,y) \tag{5.6}$$

$\bar{f}$	- ground truth, high-resolution image
g_n	- n^{th} observed low-resolution image
$\mathcal{T}_n$	- geometric transformation of nth image
h	- point spread function
$s{\downarrow}$	- down-sampling operator by a factor S
α_n, β_n	- scalar illumination parameters
η_n	- observation noise

The transformation $\mathcal{T}$ is assumed to be projective. The point spread function h is assumed to be linear, spatially invariant. The noise η is assumed to be Gaussian with mean zero.

After discretization, the model can be expressed in matrix form as

$$\mathbf{g}_n = \alpha_n \mathtt{M}_n \bar{\mathbf{f}} + \beta_n + \eta_n \tag{5.7}$$

in which the vector $\bar{\mathbf{f}}$ is a lexicographic reordering of pixels in $f(x, y)$, and where the linear operators $\mathcal{T}_n$, h and $s\downarrow$ have been combined into a single matrix $\mathtt{M}_n$, as explained in Section 5.4.4. Each low-resolution pixel is therefore a weighted sum of super-resolution pixels, the weights being determined by the registration parameters, and the shape of the point-spread function, which combines the effects of optical blur, motion blur and spatial integration.

Finally the generative models of all N images are stacked vertically to form an over-determined linear system

$$\begin{bmatrix} \mathbf{g}_0 \\ \mathbf{g}_1 \\ \vdots \\ \mathbf{g}_{N-1} \end{bmatrix} = \begin{bmatrix} \alpha_0 & & & \\ & \alpha_1 & & \\ & & \ddots & \\ & & & \alpha_{N-1} \end{bmatrix} \begin{bmatrix} \mathtt{M}_0 \\ \mathtt{M}_1 \\ \vdots \\ \mathtt{M}_{N-1} \end{bmatrix} \bar{\mathbf{f}} + \begin{bmatrix} \beta_0 \\ \beta_1 \\ \vdots \\ \beta_{N-1} \end{bmatrix} + \begin{bmatrix} \eta_0 \\ \eta_1 \\ \vdots \\ \eta_{N-1} \end{bmatrix} \tag{5.8}$$

$$\mathbf{g} = \Lambda_\alpha \mathtt{M} \bar{\mathbf{f}} + \boldsymbol{\beta} + \boldsymbol{\eta} \tag{5.9}$$

For the sake of convenience, the super-resolution pixels $\mathbf{f}$ are henceforth referred to as **super-pixels** in order to distinguish them from the low-resolution **pixels**.

It is important to note that the model described can be expressed as a purely linear transformation of the scene intensities.

5.4.3 Related approaches

Generative models found in the literature vary in their assumptions about the motion model, the point-spread function and the modelling of illumination variation. Some authors have adopted a simpler motion model, such as pure translation [10, 92, 103, 105, 113, 151, 163, 167], Euclidean [34, 52, 84] or affine [11, 85]. Others have chosen a more general motion model such as full dense stereo matching [107, 131, 136, 137]. It is also common to find other forms of point-spread function, such as a rectangular window [28], or a kernel extracted directly from calibration images [84, 97], or even from the observed image data itself [137]. At this time, it appears that no other authors have explicit photometric parameters in their model.

A small number of authors have applied super-resolution techniques to problems in which a linear relationship between scene irradiance and sensor measurement cannot be assumed, such as positron emmission tomography [125, 174].

To summarize, the model presented here is rather more general than those typically found in the literature: it allows for illumination changes, and has a quite flexible motion model.

5.4.4 Computing the elements in $\mathtt{M}_n$

In this section we address the issue of how exactly the entries in the matrices $\mathtt{M}_n$ are computed. This issue is almost universally glossed over in the literature, the only notable exceptions being the work of Smelyanskiy *et al.* [141], Zomet and Peleg [175], and Patti and Altunbasak [105].

According to the imaging model, the value of each low-resolution pixel is given by an area integral over the geometrically warped super-resolution image weighted by the point-spread function (see Figure 5.5). Each row in the matrix $\mathtt{M}$ is therefore a discretization of the integral Equation (5.6) for a single pixel.

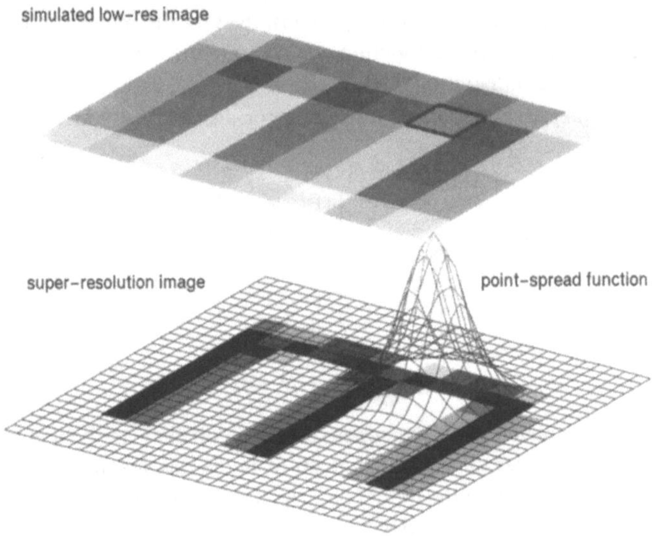

Figure 5.5: Each simulated pixel is a weighted sum of several super-pixels. The weights are determined by the geometric viewing transformation, the form and size of the point-spread function, and also by the quadrature rule used to discretize the generative model.

Essentially, there are three possible approaches to the discretization of the generative model which we shall now examine.

Method A Warp the super-resolution image into an intermediate frame using some interpolation scheme (bilinear or bicubic), convolve with a discretized version of the PSF, and finally sub-sample. The intermediate frame is aligned with the low-resolution image, but the pixel density is the same as or greater than that of the super-resolution image. This corresponds to the decomposition of $\mathtt{M}_n$ used by Zomet and Peleg [175] :

$$\mathtt{M}_n = \mathtt{DHT}_n \tag{5.10}$$

$\mathtt{D}$ - discrete down-sampling (decimation) matrix
$\mathtt{H}$ - convolution matrix formed from discretization of $h(x, y)$
$\mathtt{T}_n$ - geometric warp with suitable interpolation (bilinear/bicubic)

The matrix-vector operation $\mathtt{M}_n\mathbf{f}$ may be implemented by applying the three consecutive linear transformations to the super-resolution image. Figure 5.6 illustrates how the method applies to a one-dimensional signal. A two-dimensional schematic is shown in Figure 5.10.

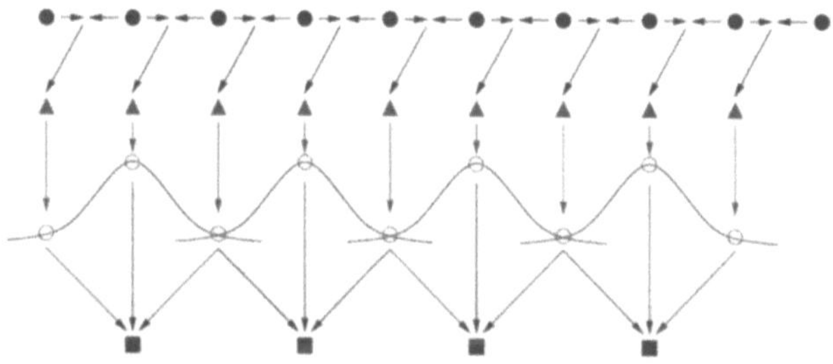

Figure 5.6: **Method A**: The super-resolution image is warped according to the geometric viewing transformation into an intermediate frame, using an interpolation scheme to compute values at non-integer positions. The intermediate frame is filtered with a discretized version of the point-spread function at positions corresponding to the low-resolution pixel centres to generate the low-resolution image. This final step combines the convolution and down-sampling operations.

Unfortunately, this scheme can give a poor approximation to the generative model, particularly in cases where perspective foreshortening is severe, resulting in large differences in scale across the image. To understand why this is, it helps to think of the image-warping process as a resampling operation, in which the resampling sites are at intra-pixel locations determined by the geometric transformation. The warping algorithm transforms pixel locations from the destination frame into the source (super-resolution) frame, and pulls back a value according to the interpolation scheme. Foreshortening effects can cause the resampling locations to be sparsely distributed in the super-resolution image, leading to under-sampling and frequency aliasing effects. Consequently, fine details in the super-resolution image may be lost or distorted even before the point-spread function has been applied. Figure 5.7 illustrates this problem.

It is, of course, possible to ameliorate this problem by increasing the resolution of the intermediate image until an acceptable sampling density is achieved in the super-resolution image. However, the associated increase in computational effort required to warp the image and perform the convolution can quickly become unacceptable.

Method B Back-project the super-pixels as discrete points (no-interpolation) into the low-resolution coordinate frame. Associate with each low-resolution pixel centre a continuous-valued PSF which is used to weight the underlying warped super-pixels. The pixel value is then the weighted sum of super-pixels (the sum of the weights having been normalized to equal 1). Figures 5.8 and 5.11 illustrate this scheme in one and two dimensions respectively.

There are two problems with this approach. Firstly, having performed the point-wise warp of the super-pixels into the low-resolution frame, determing the set of those super-pixels which lie within the "receptive field" of each truncated PSF can be hard

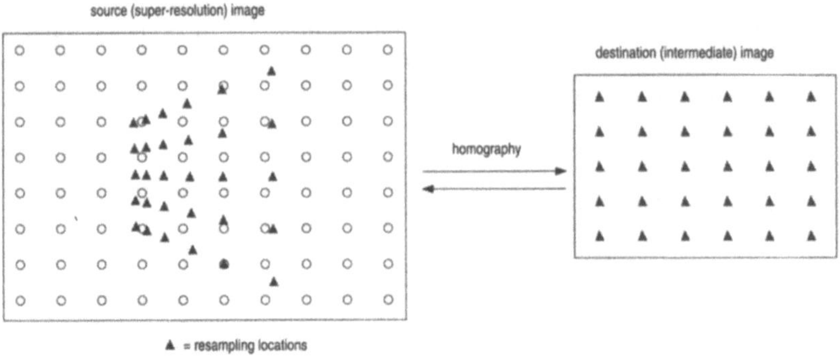

Figure 5.7: An image-warping algorithm transforms points from the destination frame into the source frame, and pulls back a value according to the interpolation scheme. Foreshortening effects can cause the resampling locations to be sparsely distributed in certain areas, leading to under-sampling and frequency aliasing effects.

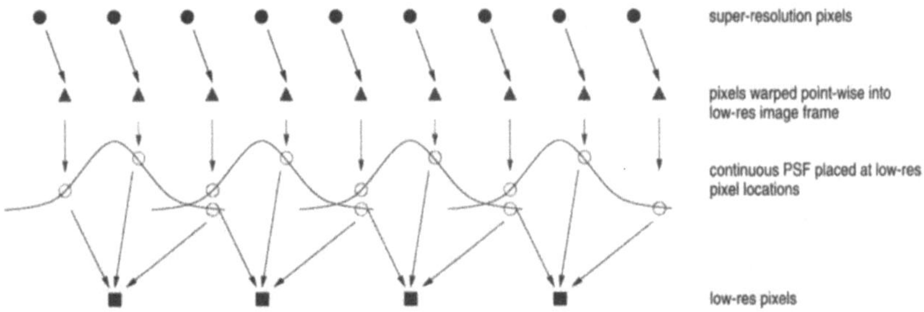

Figure 5.8: **Method B** : The super-pixels are warped point-wise into the low-resolution frame according to the geometric viewing transformation, with no interpolation. A continuous-valued PSF is associated with each low-resolution pixel centre. The PSF is truncated at, say, 3 standard deviations. Each pixel is the PSF weighted sum of super-pixels.

to implement efficiently. The problem is essentially the same as determining which of several thousand non-uniformly distributed points lies with a circle - an operation which must be repeated at every pixel.

Secondly, if the pixel-density ratio, S, is low (between 1 and 2), the receptive field will only contain a small number of super-pixels. In this case, the simple weighted sum of pixel values is not a good approximation of the required integral.

Method C Warp the continuous PSF associated with each pixel into the super-resolution frame, and making sensible assumptions about the intra-pixel continuity of the super-resolution image, evaluate the integral exactly. Figures 5.9 and 5.12 clarify the method.

The third approach is the one we use to generate the results in this book . Transformation of the Gaussian PSF of each pixel under a homography, which maps between a low-resolution image and the super-resolution image, is accomplished by computing the local affine approximation to the homography at each pixel position. Each

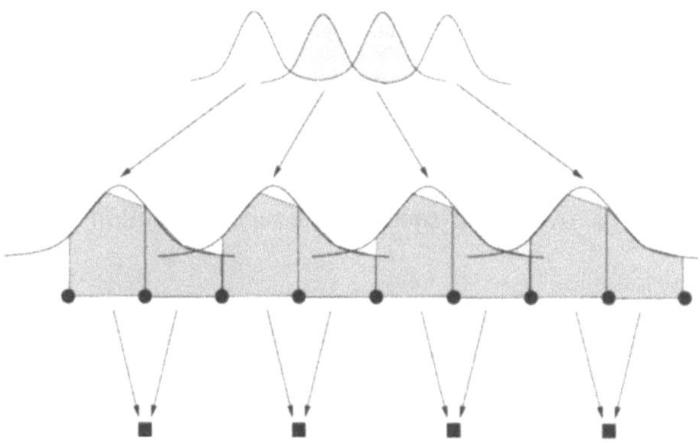

Figure 5.9: **Method C**: A continuous, truncated PSF is associated with each low-resolution pixel centre. The PSF is warped into the super-resolution frame. Both PSF and super-resolution image are assumed piece-wise bilinear, allowing the integral to be evaluated quite accurately.

Gaussian is then transformed into the super-resolution image according to the affine transformation.

Writing the geometric transformation of points (x, y) in the low-resolution image to points (x', y') in the super-resolution image as

$$(x', y') = H(x, y)$$

the required affine transformation is given by the first-order Taylor series approximation taken about a point (x_0, y_0) :

$$H(x_0 + \delta x, y_0 + \delta y) \approx H(x_0, y_0) + \nabla H \begin{bmatrix} \delta x \\ \delta y \end{bmatrix}$$

where the Jacobian ∇H is the 2×2 matrix

$$\nabla H = \begin{bmatrix} \frac{\partial H_x}{\partial x} & \frac{\partial H_x}{\partial y} \\ \frac{\partial H_y}{\partial x} & \frac{\partial H_y}{\partial y} \end{bmatrix}$$

evaluated at (x_0, y_0). Under this transformation, the parameters (μ, Σ) of the Gaussian point-spread function, $G(\mu, \Sigma)$, with centre (μ_x, μ_y) and covariance matrix Σ, are transformed into the super-resolution frame as

$$\mu' = H(\mu_x, \mu_y)$$
$$\Sigma' = (\nabla H)\, \Sigma\, (\nabla H)^\top$$

In order to compute the integral, we assume that both the PSF and the super-resolution image can be approximated by piece-wise bilinear surfaces. The value at intra-pixel positions is then given by

$$f(x + \delta x, y + \delta y) = k_{00} f_{00} + k_{10} f_{10} + k_{01} f_{01} + k_{11} f_{11}$$

where

$$
\begin{array}{llll}
f_{00} = f(x,y) & f_{10} = f(x+1,y) & f_{01} = f(x,y+1) & f_{11} = f(x+1,y+1) \\
k_{00} = (1-\delta x)(1-\delta y) & k_{10} = \delta x(1-\delta y) & k_{01} = (1-\delta x)\delta y & k_{11} = \delta x \delta y
\end{array}
$$

The integral of the surface $f \times h_{\text{psf}}$ over a unit-square in the super-resolution image is given by

$$
\begin{aligned}
\int_0^1 \int_0^1 f_{xy} h_{xy} \,\mathrm{d}x\mathrm{d}y = {} & \frac{1}{9}(f_{00}h_{00} + f_{01}h_{10} + f_{10}h_{01} + f_{11}h_{11}) \\
& + \frac{1}{18}(f_{00}(h_{10} + h_{01}) + f_{10}(h_{00} + h_{11}) \\
& + f_{01}(h_{00} + h_{11}) + f_{11}(h_{10} + h_{01})) \\
& + \frac{1}{36}(f_{00}h_{11} + f_{10}h_{01} + f_{01}h_{10} + f_{11}h_{00})
\end{aligned} \tag{5.11}
$$

In practice, the point-spread function is truncated at 3 standard-deviations, and normalized to ensure that its sum is equal to 1. The integral of $f \times h_{\text{psf}}$ over the PSF of a single pixel is computed by evaluating Equation (5.11) over every unit-square contained within the 3 standard-deviation ellipse.

Summary We can now summarize the algorithm for computing coefficients in the matrix `M`. Remember that each row in the matrix determines how a single pixel is simulated given the super-resolution image f. Each column in `M` corresponds to a particular super-pixel. The dot product of a particular row R with the discretized super-resolution image f should accurately approximate the integral of the surface $f \times h_{\text{psf}}$ over the PSF for the corresponding pixel $g(x,y)$. The coefficients on row R are determined as follows.

1. Compute the local affine transformation about the low-resolution pixel centre (x,y) into the super-resolution frame.
2. Transform the parameters of the Gaussian PSF centred on (x,y) under the affine transformation.
3. Compute the 3 standard deviation ellipse.
4. Scan-convert the ellipse to obtain the set of super-resolution pixels contained.
5. Evaluate the coefficients in Equation (5.11) for every unit-square contained within the ellipse. This implicitly defines the matrix coefficient on row R in the column C associated with each contained super-pixel $f(x,y)$.
6. Normalize the row so that its sum is equal to 1.

For problems featuring only a small number of super-pixels (say $< 100 \times 100 =$ 10 000 pixels), the coefficients for every row are pre-computed and stored in an appropriate sparse matrix structure. For larger problems, for which the memory required for explicit storage would be excessive, the matrix is implemented in a "black-box" style, with a functional interface to common operations such as left/right vector multiply. In this scheme, rows are computed on the fly as required. Caching of transformed pixel centres and transformed PSF covariance matrices is used to obtain maximum efficiency.

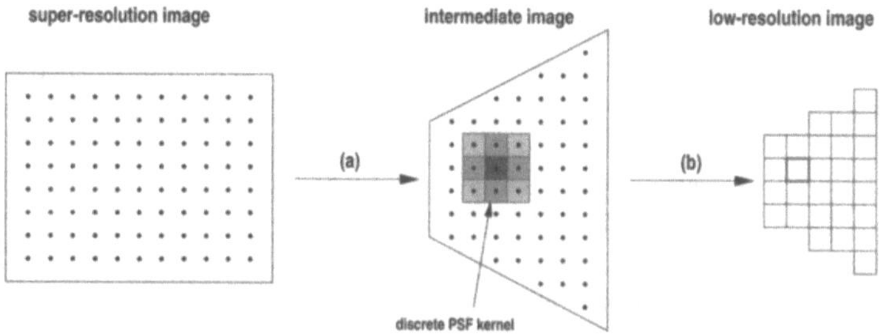

Figure 5.10: Method A: In step (a), the super-resolution image is warped according to the geometric viewing transformation into an intermediate frame, using an interpolation scheme to compute values at non-integer positions. In step (b), the intermediate frame is filtered with a discretized version of the point-spread function at positions corresponding to the low-resolution pixel centres to generate the low-resolution image.

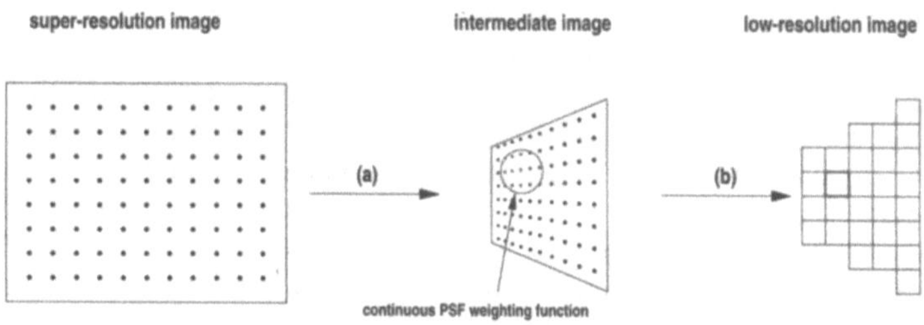

Figure 5.11: Method B: In step (a), the super-pixels are warped point-wise into the low-resolution frame according to the geometric viewing transformation, with no interpolation. In step (b), a continuous-valued PSF is associated with each low-resolution pixel centre. Each simulated pixel is a PSF weighted sum of super-pixels.

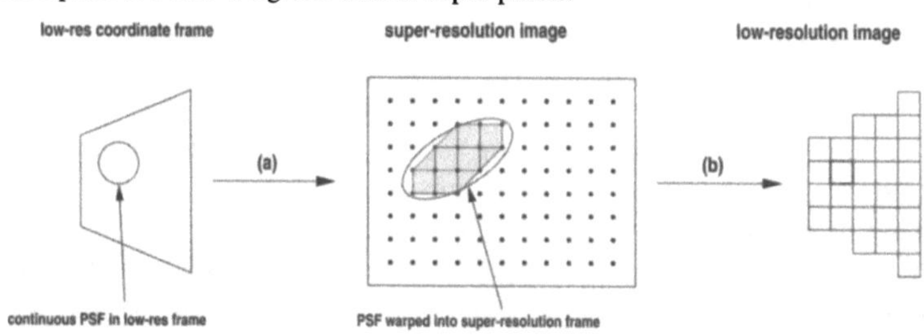

Figure 5.12: Method C: In step (a), the continuous, truncated PSF associated with each pixel centre is warped into the super-resolution frame. In step (b), both PSF and super-resolution image are assumed piece-wise bilinear, allowing the integral to be evaluated quite accurately.

5.4.5 Boundary conditions

The final implementation detail concerns the assumptions we make about the value of super-pixels which lie outside the boundary of the super-resolution image. To understand why this is necessary, consider the problem of forming the image model coefficients for a pixel which lies close to the boundary of the super-resolution image. Although the pixel centre is within the super-resolution boundary, the receptive field of its truncated PSF may extend outside. In order to simulate the value of such a pixel it is necessary to make assumptions about the values of the super-pixels beyond the boundary.

One solution to this problem is to simply discard pixels whose truncated PSF receptive field extends beyond the boundary. There are two problems with this approach. Firstly, for reasons of computational complexity, the super-resolution image is typically rather small (128×128 pixels for many of the examples shown here), and consequently the region that it covers in any low-resolution image does not constitute a very large number of pixels. Hence, the ratio of pixels near the boundary to the total number of pixels covered can be fairly significant, and so to discard these boundary pixels means throwing away a large portion of the available image data. Secondly, having discarded these pixels, there will be many super-pixels which are not in the receptive field of *any* pixels, and hence cannot be estimated. Masking out these undeterminable pixels produces a super-resolution image with a ragged boundary.

Of course, the issue of boundary conditions arises in many areas of single image restoration, denoising and deblurring. There are many common generic choices of boundary conditions, such as zero-padding, periodicity (infinite tiling of the image domain), reflection/symmetry (zero gradient across boundary), and skew-symmetry (constant gradient across boundary). In such cases, the choice of boundary conditions is usually governed by the method of solution that is to be applied, for example, use of Fourier transform techniques implies periodic boundary conditions. However, it is rare that the conditions chosen really reflect what is happening outside the image boundary.

In the multiple-view case, we can do slightly better than to assume some generic boundary conditions. The solution we adopt is to compute a crude estimate of the actual super-pixels in a border region around the super-resolution image. The width of this border is large enough to contain the PSF ellipses of all the pixels whose *centres* lie with the super-resolution boundary. The border pixels are estimated very cheaply by forming the average of the registered low-resolution images. The formation of the "average image" is covered in detail in Section 5.7, but in summary, it provides a smooth, low-noise approximation of the super-resolution image. The method is illustrated in Figure 5.13.

During any estimation procedure, the values of the border pixels remain fixed. The super-resolution image vector $\mathbf{f}$ can therefore be split into two parts, $\mathbf{f}_{\text{free}}$ and $\mathbf{f}_{\text{fixed}}$. The imaging model then becomes

$$\Lambda_\alpha \begin{bmatrix} \mathtt{M}_{\text{inside}} & \mathtt{M}_{\text{outside}} \end{bmatrix} \begin{bmatrix} \mathbf{f}_{\text{free}} \\ \mathbf{f}_{\text{fixed}} \end{bmatrix} + \boldsymbol{\beta} + \boldsymbol{\eta} = \mathbf{g}. \tag{5.12}$$

The effect of the fixed super-pixels is precomputed and subtracted from the stacked

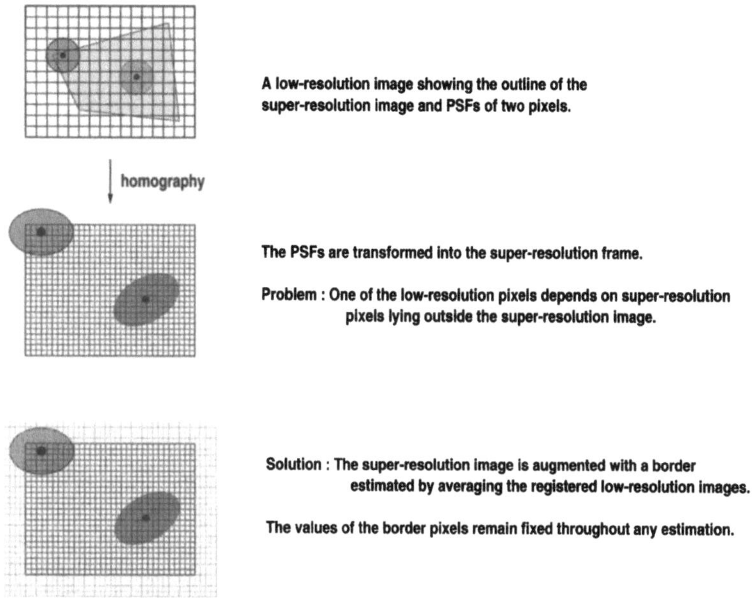

Figure 5.13: A pixel whose centre lies within the super-resolution image may have a PSF ellipse that extends beyond the boundary. To simulate such a pixel requires assumptions to be made about the values of super-pixels outside the boundary. The solution we adopt is to estimate those pixels using a cheaply computed average of the registered low-resolution images.

low-resolution image vector g prior to beginning any estimation procedure :

$$\begin{aligned} \Lambda_\alpha(\mathtt{M}_{\text{inside}}\mathbf{f}_{\text{free}} + \mathtt{M}_{\text{outside}}\mathbf{f}_{\text{fixed}}) + \boldsymbol{\beta} + \boldsymbol{\eta} &= \mathbf{g} \\ \Rightarrow \Lambda_\alpha\mathtt{M}_{\text{inside}}\mathbf{f}_{\text{free}} + \boldsymbol{\beta} + \boldsymbol{\eta} &= \mathbf{g} - \Lambda_\alpha\mathtt{M}_{\text{outside}}\mathbf{f}_{\text{fixed}} \end{aligned} \tag{5.13}$$

5.5 Justification for the Gaussian PSF

In this section we perform a simple experiment to measure a cross-section through the PSF of a real camera. The method is a modification of that proposed by Reichenbach *et al.* [117]. A sharp white-black step edge is printed using a laser printer. The edge is imaged with a monochrome Pulnix CCD camera arranged so that there is a very acute (but not zero) angle between the direction of the edge and the horizontal scan lines of the camera. Canny edges are extracted in the captured image, and a line is fitted to them by orthogonal regression. This provides a very accurate estimate of the sub-pixel location of the step edge. The distance of any image pixel to the step edge may then be calculated to sub-pixel accuracy. By considering all pixels that lie within a few pixels of the edge, we can obtain a very dense super-sampling of the transition.

The convolution of the PSF with a step edge produces the cumulative integral of the PSF. Under our Gaussian PSF assumption we would therefore expect the edge

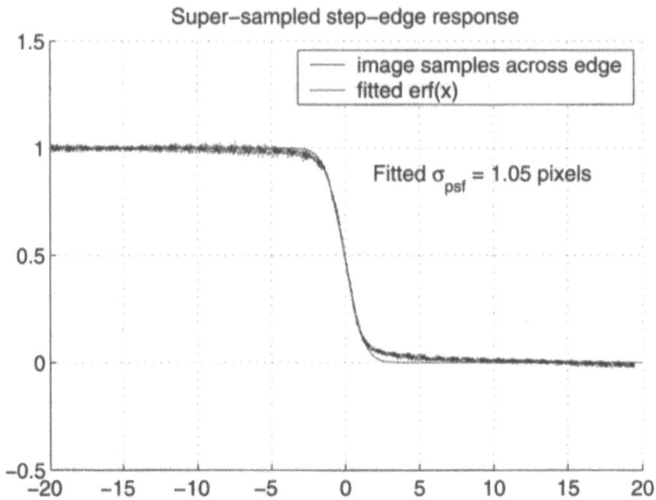

Figure 5.14: The super-sampled edge response of a Pulnix CCD camera imaging a sharp white-black transition. The best-fitting $\mathrm{erf}(x)$ is overlaid, corresponding to a Gaussian PSF with $\sigma = 1.05$ pixels. The close fit is evidence that the Gaussian approximation is reasonable.

response to correspond to the *error function* $\mathrm{erf}(x)$. Figure 5.14 shows the super-sampled edge response. The best-fitting $\mathrm{erf}(x)$ is overlaid. This corresponds to a Gaussian PSF with $\sigma = 1.05$ pixels. The fit is fairly close, evidence that the Gaussian approximation is quite reasonable.

5.6 Synthetic test images

A principal aim of this research is to develop algorithms which work on real image sequences. However, on several occasions throughout this and later chapters we will make use of synthetically generated test sequences. This has two benefits over using only real data :

- Comparisons can be made between the known ground-truth image and the estimated super-resolution image.
- Factors of the image model which affect the super-resolution estimate can be controlled and investigated in isolation. Such factors include the accuracy of registration, accuracy of the point-spread function, and level of observation noise.

Figure 5.15 shows the four ground-truth images from which all the synthetic sequences are generated. The images are all 128×128 pixel, 8-bit gray-scale images. Each image exhibits different characteristics which are of interest :

- **"Text"** is a high-contrast image with fine detail.

- **"Test card"** is bi-level image, with bar features of varying spatial frequency.
- **"Faraday"** is an extract from a bank note, in which the grey-levels are coarsely quantized. The image structure has a fairly piece-wise constant profile.
- **"Face"** exhibits both areas of detail and areas of smooth variation.

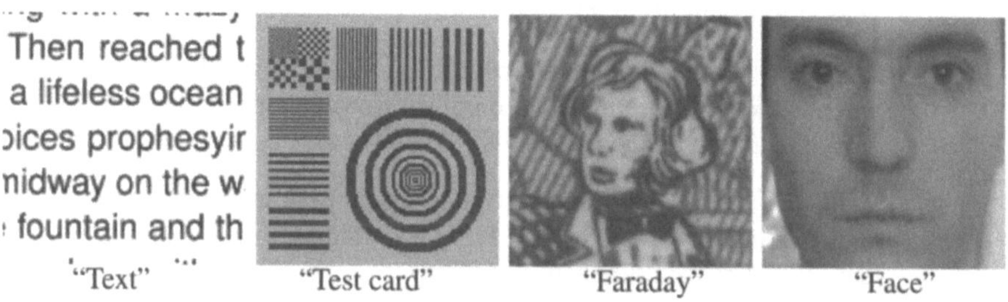

Figure 5.15: The four ground truth images used to generate the synthetic sequences used in this book .

The method by which the synthetic sequences are generated is outlined in Table 5.1 . The synthesized images are also 8-bit grey-level images and have half the pixel density of the ground-truth images.

Generating the random homographies In order to synthesize a variety of different views it is necessary to generate a set of random planar homographies, each of which maps coordinates in the super-resolution frame into the coordinate frame of a particular low-resolution image. Simply randomizing the homography matrix elements does not give satisfactory results. Instead, we use an algorithm which mimicks the real image formation process. The super-resolution image is modelled as a unit-square planar surface which is rotated and translated in space and viewed by a synthetic pinhole camera. The procedure is outlined in Table 5.2 , and also in Figure 5.16.

The first five synthetic images generated from each ground-truth image are shown in Figure 5.17. For examples based on real images, the data sets will be introduced alongside each example.

5.7 The average image

The average image is formed by a simple resampling scheme applied to the registered input images. The image formed is typically an overly smooth approximation to the super-resolution image, which is extremely robust to noise in the observed images. As such, it provides an excellent starting point for any super-resolution estimation procedure. As explained in Section 5.4.5, it also provides a reasonable estimate of super-pixels outside the boundary of the super-resolution image, and is hence used to provide Dirichlet boundary conditions.

The imaging model described in the previous sections encodes how a particular low-resolution pixel g_j is simulated by forming a weighted sum over the set of super-pixels $\{f_i\}$ with weights $\{w_i\}$. Only super-pixels contained within the pixel's PSF

Objective Generate synthetic low-resolution images.

Algorithm

1. Generate a set of 50 randomly chosen homographies (this procedure is detailed in the text).

2. Warp the ground-truth image according to the homographies.

3. Convolve the resulting images with a Gaussian with standard deviation 2 pixels. This simulates combined isotropic optical blur and area integration by the CCD cells.

4. Scale the image intensities by a random factor chosen from a normal distribution, $\alpha \sim \mathcal{N}(1, 0.1)$. This simulates the "contrast" component of the illumination model.

5. Shift the image intensities by a random amount chosen from a normal distribution, $\beta \sim \mathcal{N}(0, 5)$. This simulates the "brightness" part of the illumination model.

6. Down-sample by a factor of 2. This simulates the spatial quantization due to the digitization process.

Table 5.1: The main steps in the procedure by which synthetic low-resolution images are generated.

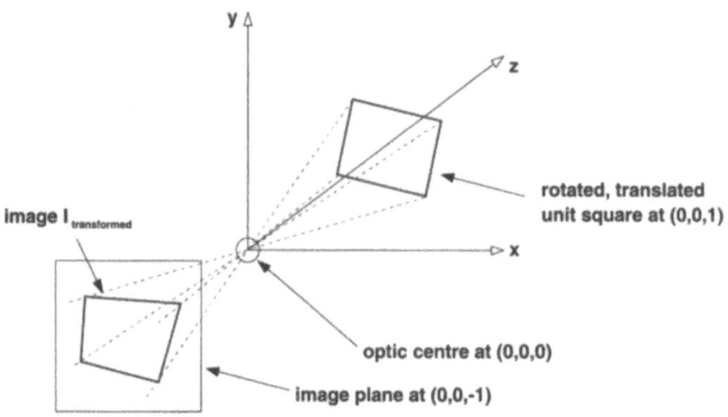

Figure 5.16: A simple synthetic pin-hole camera viewing a unit-square, which is rotated and translated in space, is used to generate random homographies.

ellipse (receptive field) have non-zero w_i. The "average image" is formed by considering the adjoint operation, i.e. f_i^{avg} is formed from a weighted sum of those pixels which depend on f_i. To clarify :

1. For each super-pixel f_i, find, in all low-resolution images, the set of pixels

Objective Generate synthetic projective image transformations.
Algorithm
1. Position a synthetic pin-hole camera with unit focal length and aspect ratio at the origin, the optic axis aligned with the z-axis.
2. Position a unit-square directly in front of the camera, parallel to the image plane and unit distance from the origin. The projection of the unit-square onto the image plane forms an image which is also a unit-square, denoted as I_{square}.
3. Rotate the unit square about the x, y and z axes by randomly chosen values in the range ± 15 degrees.
4. Translate the unit square parallel to the image plane by randomly chosen x and y distances in the range ± 0.01.
5. Project the transformed square onto the image plane. Denote this image as $I_{\text{transformed}}$.
6. Compute the homography which maps I_{square} to $I_{\text{transformed}}$.
7. Post-multiply by a similarity transformation mapping the required super-resolution coordinate frame to the unit square.
8. Pre-multiply by a similarity transformation mapping the unit-square to the required low-resolution coordinate frame.

Table 5.2: The algorithm used to generate realistic projective image transformations.

$\{g_j\}$ that depend on f_i with weight w_j (i.e. those that contain f_i in their PSF receptive field.)

2. Compute $f_i^{\text{avg}} = \frac{\sum_j w_j g_j}{\sum_j w_j}$.

This is easily expressed in terms of the image model matrix as

$$\mathbf{f}_{\text{avg}} = \mathtt{K}^{-1}\mathtt{M}^{\top}\Lambda_{\alpha}^{-1}(\mathbf{g} - \boldsymbol{\beta}) \tag{5.14}$$

where $\mathtt{K}$ is a diagonal matrix whose non-zero entries are the column sums of $\mathtt{M}$,

$$\mathtt{K}_{ii} = \sum_{\forall j} \mathtt{M}_{ji} \tag{5.15}$$

Note that, in the above formulation, the low-resolution images have been pre-adjusted to normalize their illumination parameters. Figure 5.18 shows the average image formed from our four synthetic sequences.

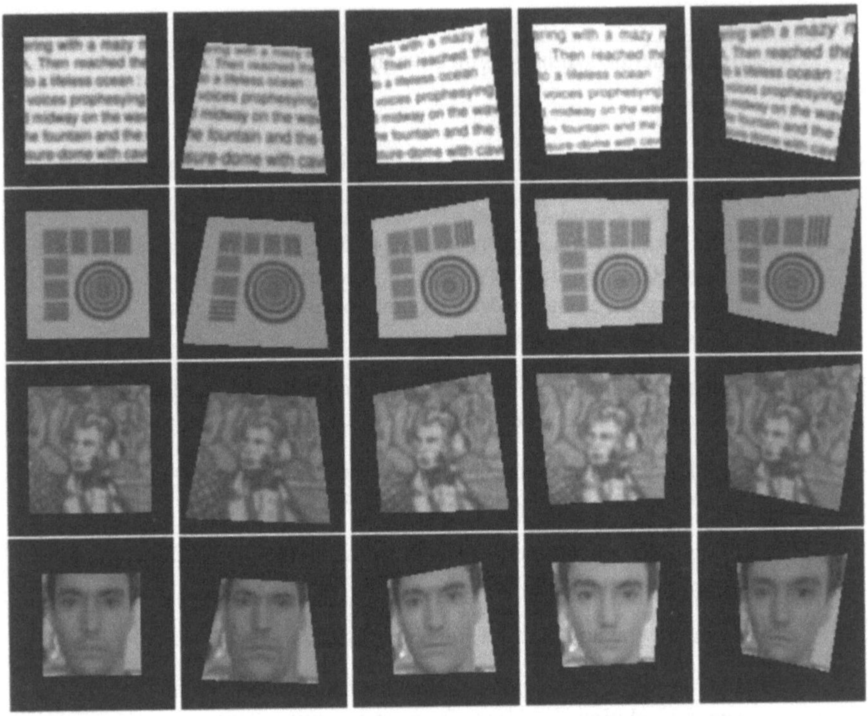

Figure 5.17: The first five synthetic images generated from each ground-truth image.

Figure 5.18: (Top) The ground-truth images. (Bottom) The average images formed from each of the four synthetic low-resolution image sequences. The average image is a good starting point for estimation of the super-resolution image.

5.7.1 Noise robustness

As we might expect, the average image is extremely robust to additive noise in the low-resolution images. This fact is clearly demonstrated in Figure 5.19, in which an average image is formed from just ten low-resolution images which have been corrupted by extremely large levels of additive Gaussian noise.

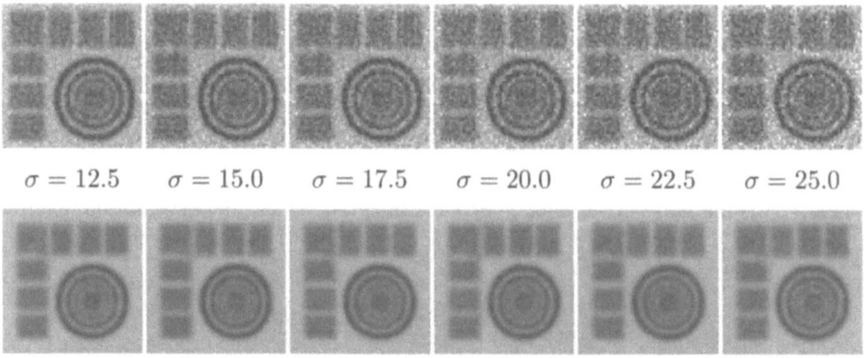

Figure 5.19: Average images formed from 10 synthetic low-resolution images from the "Test card" sequence, with increasing levels of additive Gaussian noise corrupting the images. (Top) One of the low-resolution images with noise standard deviation σ grey-levels. (Bottom) The corresponding average image for each noise level.

In fact, for a given number of input images N, each contaminated by spatially invariant, mean zero, additive noise with variance σ^2, the expected noise variance in the average image can be quite accurately estimated, as we shall now demonstrate.

Denoting a noise-free low-resolution image as $\bar{\mathbf{g}}_n$, and the average image formed by a combination of noise-free images as $\bar{\mathbf{f}}_{\text{avg}}$, the nth noisy, low-resolution image can be written as

$$\mathbf{g}_n = \bar{\mathbf{g}}_n + \boldsymbol{\eta}_n$$

Hence,

$$\begin{aligned}\mathbf{f}_{\text{avg}} &= \mathtt{K}^{-1}\mathtt{M}^{\top}\Lambda_{\alpha}^{-1}(\bar{\mathbf{g}} + \boldsymbol{\eta} - \boldsymbol{\beta}) \\ &= \mathtt{K}^{-1}\sum_{\forall n}\alpha_n^{-1}\mathtt{M}_n^{\top}(\bar{\mathbf{g}} + \boldsymbol{\eta}_n - \beta_n)\end{aligned}$$

The expectation $\mathrm{E}(\mathbf{f}_{\text{avg}})$ is therefore

$$\begin{aligned}\mathrm{E}(\mathbf{f}_{\text{avg}}) &= \mathrm{E}(\mathtt{K}^{-1}\sum_{\forall n}\alpha_n^{-1}\mathtt{M}_n^{\top}(\bar{\mathbf{g}} + \boldsymbol{\eta}_n - \beta_n)) \\ &= \mathtt{K}^{-1}\sum_{\forall n}\alpha_n^{-1}\mathtt{M}_n^{\top}(\mathrm{E}(\bar{\mathbf{g}}) + \mathrm{E}(\boldsymbol{\eta}_n) - \mathrm{E}(\beta_n)) \\ &= \mathtt{K}^{-1}\sum_{\forall n}\alpha_n^{-1}\mathtt{M}_n^{\top}(\bar{\mathbf{g}} - \beta_n) \\ &= \bar{\mathbf{f}}_{\text{avg}}\end{aligned}$$

Denoting the error as $\mathbf{e}_{\text{avg}} = \mathbf{f}_{\text{avg}} - \bar{\mathbf{f}}_{\text{avg}}$, the variance $\Sigma(\mathbf{e}_{\text{avg}})$ is given by

$$\begin{aligned}
\Sigma(\mathbf{e}_{\text{avg}}) &= \text{Var}[\mathtt{K}^{-1} \sum_{\forall n} \alpha_n^{-1} \mathtt{M}_n^{\top} (\bar{\mathbf{g}} + \eta_n - \beta_n)] \\
&= \mathtt{K}^{-1} (\sum_{\forall n} \alpha_n^{-2} \mathtt{M}_n^{\top} \, \Sigma[\bar{\mathbf{g}} + \eta_n - \beta_n] \, \mathtt{M}_n) \mathtt{K}^{-1} \\
&= \mathtt{K}^{-1} (\sum_{\forall n} \alpha_n^{-2} \mathtt{M}_n^{\top} \, \Sigma_{\eta_n} \, \mathtt{M}_n) \mathtt{K}^{-1}
\end{aligned}$$

where Σ_{η_n} is the noise covariance matrix for image n. Under the assumption that the noise is spatially uncorrelated, with variance σ^2 being the same in every image, the equation becomes

$$\begin{aligned}
\Sigma(\mathbf{e}_{\text{avg}}) &= \mathtt{K}^{-1} (\sum_{\forall n} \alpha_n^{-2} \sigma^2 \mathtt{M}_n^{\top} \mathtt{M}_n) \mathtt{K}^{-1} \\
&= \sigma^2 \mathtt{K}^{-1} (\sum_{\forall n} \alpha_n^{-2} \mathtt{M}_n^{\top} \mathtt{M}_n) \mathtt{K}^{-1}
\end{aligned} \tag{5.16}$$

Now, Equation (5.15) for the diagonal matrix of column sums, $\mathtt{K}$, may be re-written in terms of the column-sums of each individual image model matrix $\mathtt{M}_n$:

$$\begin{aligned}
\mathtt{K} &= \sum_{\forall n} \mathtt{K}_n \\
\mathtt{K}_{n_{ii}} &= \sum_{\forall j} \mathtt{M}_{n_{ji}}
\end{aligned}$$

In general, there is very little variation among the values of the column sums for a particular image model matrix $\mathtt{M}_n$, and we can therefore assume that $\mathtt{K}_n = k_n \mathtt{I}$, a multiple of the identity. Hence,

$$\mathtt{K} \approx \sum_{\forall n} k_n \mathtt{I}$$

for some constant k_n, the average column sum of $\mathtt{M}_n$. Furthermore, within any particular sequence of N images there is typically very little variation between the column sums of different image model matrices, allowing the additional simplification of assuming that $\mathtt{K}$ is directly proportional to the number of images N :

$$\mathtt{K} \approx kN\mathtt{I} \tag{5.17}$$

where k is the average k_n over all n.

These assumptions are verified in Figure 5.20, which shows histograms of the column sums for five different image model matrices $\mathtt{M}_n$. Under these assumptions, Equation (5.16) reduces to

$$\begin{aligned}
\Sigma(\mathbf{e}_{\text{avg}}) &= \sigma^2 (kN\mathtt{I})^{-1} (\sum_{\forall n} \alpha_n^{-2} \mathtt{M}_n^{\top} \mathtt{M}_n) (kN\mathtt{I})^{-1} \\
&= \frac{\sigma^2}{(kN)^2} \sum_{\forall n} \alpha_n^{-2} \mathtt{M}_n^{\top} \mathtt{M}_n
\end{aligned}$$

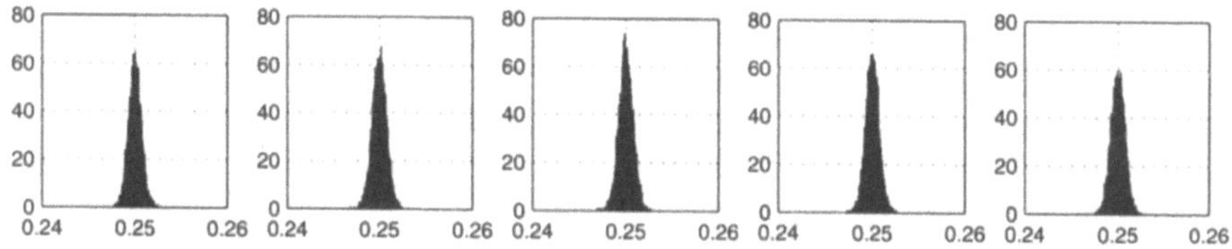

Figure 5.20: Histograms of the columns sums for five different image model matrices $\mathtt{M}_n$ from the "Test card" synthetic sequence. There is little variation either within or between the different images.

Given that we are interested in the typical variance of a single pixel in the average image, we can safely ignore the off-diagonal elements of the covariance matrix for now, leaving

$$\mathrm{Var}(\mathbf{e}_{\mathrm{avg}}) = \frac{\sigma^2}{(kN)^2} \sum_{\forall n} \alpha_n^{-2} \mathrm{diag}(\mathtt{M}_n^{\top} \mathtt{M}_n) \tag{5.18}$$

We can now make our first useful observation: for a fixed number of images, we can expect the noise variance in the average image to be directly proportional to the noise variance in the low-resolution images. Indeed, this relationship is verified empirically in Figure 5.22.

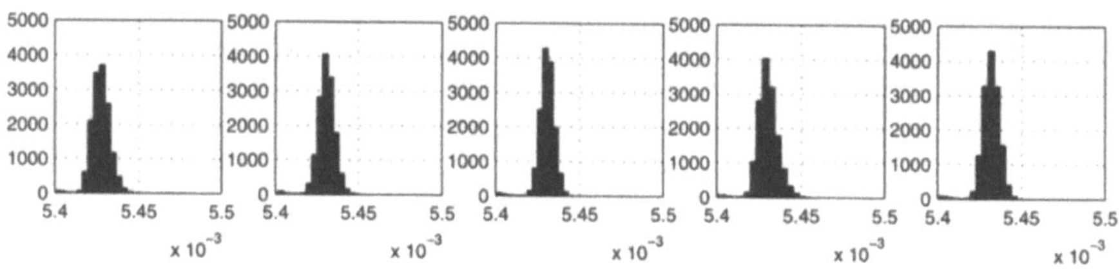

Figure 5.21: Histograms of the diagonal elements of $\mathtt{M}_n^{\top}\mathtt{M}_n$ for five different image model matrices $\mathtt{M}_n$. There is little variation either within or between the different images.

Considering the diagonal elements of a particular matrix $\mathtt{M}_n^{\top}\mathtt{M}_n$, we can again assume that there is little variation among these elements. Neither is there significant variation of these elements between different $\mathtt{M}_n^{\top}\mathtt{M}_n$. This fact is verified in Figure 5.21, which shows histograms of the diagonal elements of $\mathtt{M}_n^{\top}\mathtt{M}_n$ for five different images. We can therefore make another simplification :

$$\sum_{\forall n} \mathrm{diag}(\mathtt{M}_n^{\top} \mathtt{M}_n) \approx dN\mathtt{I} \tag{5.19}$$

for some constant d, the average of the diagonal elements of $\mathtt{M}_n^{\top}\mathtt{M}_n$ over all n. Equation (5.18) then becomes

$$\begin{aligned} \mathrm{Var}(\mathbf{e}_{\mathrm{avg}}) &= \frac{\sigma^2}{(kN)^2} dN \sum_{\forall n} \alpha_n^{-2} \mathtt{I} \\ &= \frac{d}{k^2} \frac{\sigma^2}{N} \sum_{\forall n} \alpha_n^{-2} \mathtt{I} \end{aligned} \tag{5.20}$$

The value of the photometric parameter α_n is typically very close to one, and hence the summation $\sum_{\forall n} \alpha_n^{-2}$ is generally near unity. This leads to our second observation : for fixed image noise variance σ^2, we expect the noise variance in the average image to be inversely proportional to the number of low-resolution images available. This relationship is also verified empirically in Figure 5.23.

Finally, let us consider the ratio $\frac{d}{k^2}$. Both constants depend on the number of columns in the image model matrices $\mathtt{M}_n$. The number of columns is in turn proportional to S^2, the square of the pixel-zoom ratio. Remember that the row sums of $\mathtt{M}_n$ are always equal to one, hence no matter how many columns there are, the total sum of elements in $\mathtt{M}_n$ is constant (the number of rows is equal to the number of pixels, and hence fixed). Consequently, from the definition of k as the average column sum of $\mathtt{M}$ we obtain the relationship

$$k \approx \frac{1}{NS^2}$$

Each diagonal element of $\mathtt{M}^\top \mathtt{M}$ is simply the squared two-norm of the corresponding column of $\mathtt{M}$. Remember that the non-zero elements in a particular column indicate the contribution that the corresponding super-pixel makes to a number of different low-resolution pixels (those which include the super-pixel in their receptive fields.) In general, and given a large number of images, the coefficients in a column will uniformly sample the point-spread function, and their sum will of course be equal to the column sum. Hence, the two-norm of a column is equal to the two-norm of the point-spread function multiplied by the column sum. So, from the definition of d as the average of the diagonal elements of $\mathtt{M}^\top \mathtt{M}$, we obtain :

$$d \approx k^2 ||h_{\text{psf}}||^2$$

Hence

$$\frac{d}{k^2} = ||h_{\text{psf}}||^2$$

In the case of a Gaussian PSF, as here, we can easily obtain an analytic expression for the squared two-norm of h_{psf}, and hence finally obtain

$$\frac{d}{k^2} = \frac{1}{2\sigma_{\text{psf}}\sqrt{\pi}}$$

Note that, although the pixel-zoom S appears in the expression for k, it does not affect the ratio $\frac{d}{k^2}$. Thus we arrive at our final pair of observations: the ratio $\frac{d}{k^2}$ is directly proportional to the squared magnitude of the point-spread function, and invariant to the pixel-zoom ratio S. This final result is demonstrated empirically in Figures 5.24 and 5.25.

To summarize, we have the following results concerning the noise variance $\text{Var}(\mathsf{e}_{\text{avg}})$ in the average image :

1. $\text{Var}(\mathsf{e}_{\text{avg}})$ is directly proportional to σ^2, the noise variance in the low-resolution images.

2. $\text{Var}(\mathsf{e}_{\text{avg}})$ is inversely proportional to the number of low-resolution images used.

3. $\mathrm{Var}(\mathbf{e}_{\mathrm{avg}})$ is directly proportional to the squared magnitude of the point-spread function $\|\sigma_{\mathrm{psf}}\|^2$.

4. $\mathrm{Var}(\mathbf{e}_{\mathrm{avg}})$ is invariant to the pixel zoom ratio, S.

Combining these results we obtain

$$\mathrm{Var}(\mathbf{e}_{\mathrm{avg}}) \approx \frac{\sigma^2}{2N\sigma_{\mathrm{psf}}\sqrt{\pi}} \tag{5.21}$$

The importance of this result will become clearer in Section 5.9, where we consider the properties of the maximum-likelihood super-resolution estimator.

The graphs in figures 5.22 to 5.25 were generated by computing the average image using synthetic images from the "Test card" sequence, with Gaussian noise of the specified variance added to the synthetic images. The noise variance $\mathrm{Var}(\mathbf{e}_{\mathrm{avg}})$ is computed by comparison with the average image formed from noise-free images, $\mathbf{f}_{\mathrm{avg}}$.

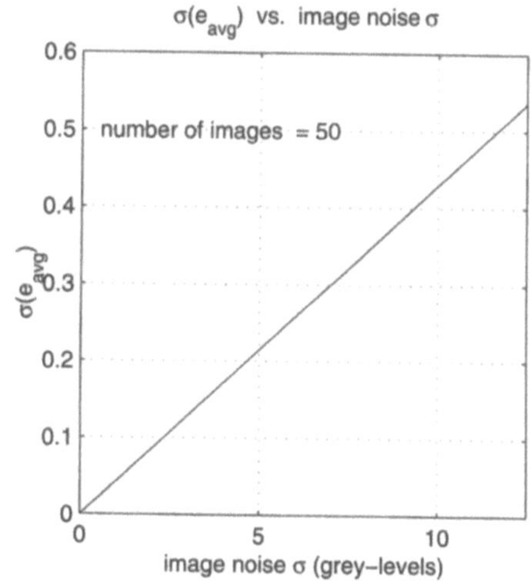

Figure 5.22: $\mathrm{Var}(\mathbf{e}_{\mathrm{avg}})$ is plotted against σ^2 for increasing levels of additive Gaussian noise. The number of low-resolution images is fixed ($N = 50$). The graph demonstrates the direct proportionality between $\mathrm{Var}(\mathbf{e}_{\mathrm{avg}})$ and σ^2.

5.8 Rudin's forward-projection method

In this section, we briefly review the super-resolution estimator proposed by Rudin *et al.* [122]. The algorithm is quite simple and generally produces good results. We derive a result which gives some insight into its success, and explain the conditions under which it is necessary to employ the more computationally expensive techniques proposed in this book. The proposed method is as follows.

1. Pre-compute the dense point-to-point correspondence between views.

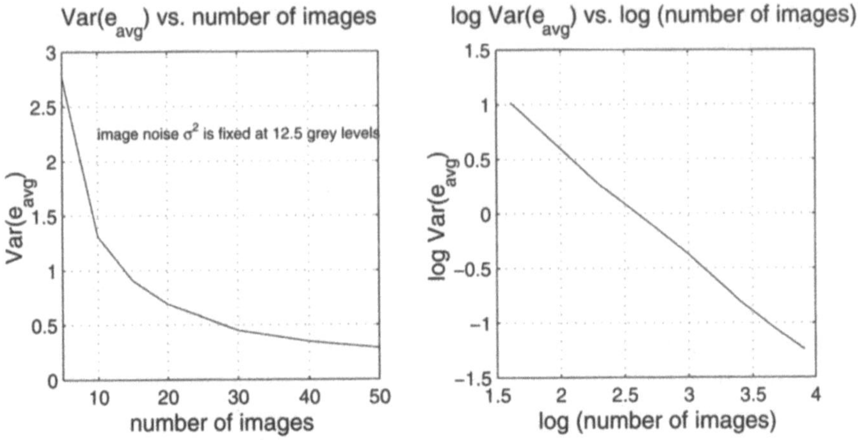

Figure 5.23: (Left) $\text{Var}(\mathbf{e}_{\text{avg}})$ is plotted against the number of low-resolution images used. The image noise variance is fixed ($\sigma^2 = 12.5$ grey-levels). (Right) When plotted on log axes, the graph clearly demonstrates the inverse proportionality between $\text{Var}(\mathbf{e}_{\text{avg}})$ and N.

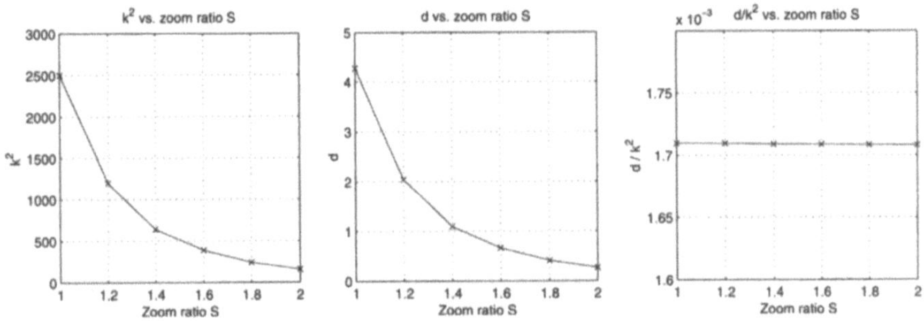

Figure 5.24: These three graphs demonstrate that, although there is significant variation in d and k^2 with respect to the pixel zoom ratio, S, the ratio $\frac{d}{k^2}$ is nearly constant.

2. Choose one input image as a reference coordinate frame.

3. For each input image, *forward warp* every pixel's *coordinates* into the reference frame (i.e. no interpolation at this stage).

4. Resample the resulting irregularly spaced data onto a uniform grid of pixels at high spatial density. Each resampling location takes a weighted average of nearby samples, weighted according to distance ($w = e^{-|\delta x|}$).

5. Finally, deblur the resulting image using a standard single-image method [33, 169].

Figure 5.26 illustrates the process.

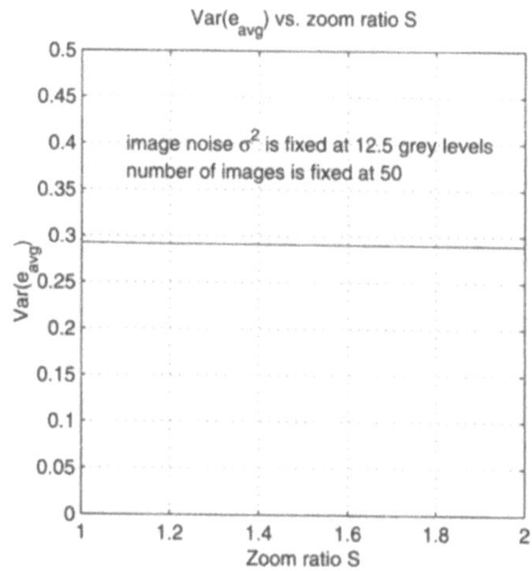

Figure 5.25: $\text{Var}(\mathbf{e}_{\text{avg}})$ is plotted against pixel zoom ratio S. The number of images and noise variance are fixed ($N = 50$, $\sigma^2 = 12.5$ grey-levels). $\text{Var}(\mathbf{e}_{\text{avg}})$ is effectively invariant with respect to S.

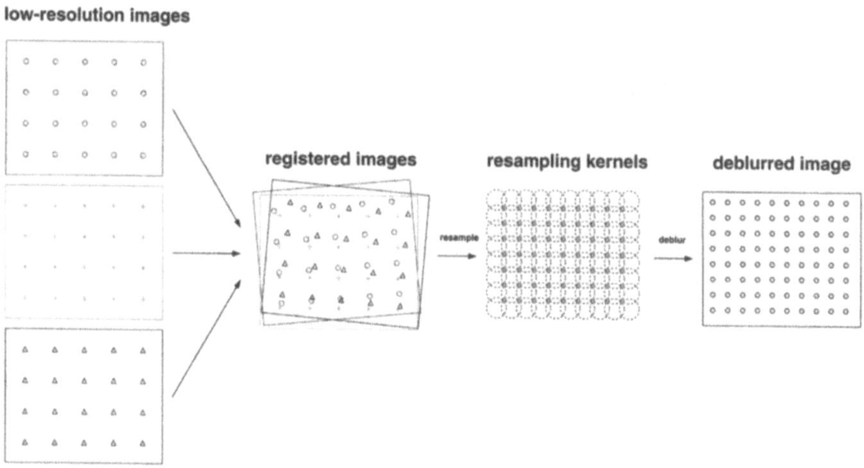

Figure 5.26: Schematic representation of Rudin's forward-projection super-resolution algorithm.

The authors choose to employ an expensive dense-stereo matching algorithm to obtain image registration. This is purely a detail of their specific implementation however, and in general any accurate registration algorithm appropriate for the particular motion model may be substituted without changing the essence of the method.

Figure 5.28 shows the result of applying the algorithm to 50 synthetic images generated from the ground-truth "Test card" image. The images differ from those shown in Figure 5.17 in that the image-to-image transformations are only Euclidean (translation and rotation) and there is no simulation of illumination changes (see Figure 5.27).

The reason for this will be become clear shortly.

The reconstructed image is reasonably good, although it does not capture the very finest details. The relative error[1] is 0.18, equivalent to an rms error of 30.7 grey-levels.

Figure 5.27: Five of the 50 synthetic images generated from the "Test card" ground-truth image for the purpose of demonstrating Rudin's forward-projection algorithm. Unlike those in Figure 5.17, the image-to-image transformations are purely Euclidean (translation/rotation only).

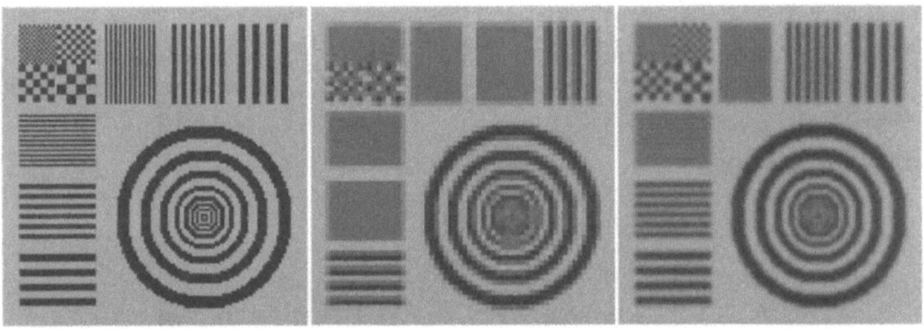

Figure 5.28: The result of applying Rudin's forward projection method to 50 synthetic images, a sample of which are shown in Figure 5.27. (Left) The ground-truth image. (Middle) One of the low-resolution input images. (Right) The reconstructed high-resolution image.

To understand the success of the method, we consider the following component-wise decomposition of the imaging process (the effects of illumination variation can safely be ignored for the moment)

$$g_n = s{\downarrow}(h * \mathcal{T}_n(f)) \tag{5.22}$$

If the point-spread function h is *isotropic*, and if the viewing transformation $\mathcal{T}_n$ is *Euclidean*, then the operations of geometric warping and convolution with the PSF are *commutable*, so

$$g_n = s{\downarrow}(\mathcal{T}_n(h * f)) \tag{5.23}$$

In other words, it makes no difference whether blurring occurs prior to warping, or post-warping. What this means is that, under these restricted conditions, a perfectly valid way to obtain a super-resolution estimate from the low-resolution images is to first reverse the down-sampling (i.e. up-sample), then to undo the geometric distortion of each frame (i.e register and resample), and finally to apply a single deblurring

[1]The relative error between the ground-truth $\mathbf{f}$ and the reconstruction $\hat{\mathbf{f}}$ is defined as the ratio $\frac{|\mathbf{f}-\hat{\mathbf{f}}|_2}{|\mathbf{f}|_2}$.

step to the combined registered/resampled data. This is essentially the method Rudin proposes.

In many real-life situations, the Euclidean motion model is fairly accurate, e.g. small objects, distant from the camera moving parallel to the image plane; or a small region-of-interest in a small-angle camera pan. In these situations, Rudin's method can perform well. However, as has already been noted, the method cannot recapture the finest details in the super-resolution image. This is probably due to the somewhat unsatisfactory method which is used to resample a band-limited signal encoded by a set of irregularly spaced samples. The exponential distance weighting method adopted by Rudin is only a heuristic solution to this problem. A superior solution would involve an optimization problem in which an estimate is obtained of the uniformly spaced samples which, when optimally interpolated using a sinc kernel, best approximate the irregularly spaced data.

5.9 The maximum-likelihood estimator

In this section we derive a maximum-likelihood estimator for the super-resolution image. Assuming the image noise to be Gaussian with mean zero, variance σ^2, the total probability of an observed image $g_n(x,y)$ given an estimate of the super-resolution image $\hat{f}(x,y)$ is

$$\Pr(\mathbf{g}_n|\hat{\mathbf{f}}) = \prod_{\forall x,y} \frac{1}{\sigma\sqrt{2\pi}} \exp\left(-\frac{(\hat{g}_n(x,y)-g_n(x,y))^2}{2\sigma^2}\right) \tag{5.24}$$

where the simulated low-resolution image $\hat{\mathbf{g}}_n$ is given by

$$\hat{\mathbf{g}}_n = \alpha_n \mathtt{M}_n \hat{\mathbf{f}} + \beta_n$$

The corresponding log-likelihood function is

$$\begin{aligned} \mathcal{L}(\mathbf{g}_n) &= -\sum_{\forall x,y} (\hat{g}_n(x,y)-g_n(x,y))^2 \\ &= -\|\hat{\mathbf{g}}_n - \mathbf{g}_n\|^2 \\ &= -\|\alpha_n \mathtt{M}_n \hat{\mathbf{f}} + \beta_n - \mathbf{g}_n\|^2 \end{aligned} \tag{5.25}$$

The log-likelihood over all images is given by

$$\begin{aligned} \sum_{\forall n} \mathcal{L}(\mathbf{g}_n) &= -\sum_{\forall n} \|\alpha_n \mathtt{M}_n \hat{\mathbf{f}} + \beta_n - \mathbf{g}_n\|^2 \\ &= -\|\Lambda_\alpha \mathtt{M}\hat{\mathbf{f}} + \boldsymbol{\beta} - \mathbf{g}\|^2 \end{aligned}$$

We seek an estimate $\mathbf{f}_{\text{mle}}$ which maximizes the log-likelihood over all images :

$$\begin{aligned} \mathbf{f}_{\text{mle}} &= \arg\max_{\mathbf{f}} \sum_{\forall n} \mathcal{L}(\mathbf{g}_n) \\ &= \arg\max_{\mathbf{f}} \left(-\|\Lambda_\alpha \mathtt{M}\mathbf{f} + \boldsymbol{\beta} - \mathbf{g}\|^2\right) \\ &= \arg\min_{\mathbf{f}} \|\Lambda_\alpha \mathtt{M}\mathbf{f} + \boldsymbol{\beta} - \mathbf{g}\|^2 \end{aligned} \tag{5.26}$$

The optimum $\mathbf{f}_{\text{mle}}$ is given by

$$\begin{aligned} \frac{\mathrm{d}\mathcal{L}}{\mathrm{d}\mathbf{f}} &= 2\mathtt{M}^\top \Lambda_\alpha^\top(\Lambda_\alpha \mathtt{M}\mathbf{f} + \boldsymbol{\beta} - \mathbf{g}) = 0 \\ \Rightarrow \hat{\mathbf{f}}_{\text{mle}} &= (\mathtt{M}^\top \Lambda_\alpha{}^2 \mathtt{M})^{-1} \mathtt{M}^\top \Lambda_\alpha (\mathbf{g} - \boldsymbol{\beta}) \\ &= (\Lambda_\alpha \mathtt{M})^+ (\mathbf{g} - \boldsymbol{\beta}) \end{aligned} \tag{5.27}$$

where $(\Lambda_\alpha \mathtt{M})^+$ is a Moore–Penrose inverse.

We now show that $\mathbf{f}_{\text{mle}}$ is an unbiased estimator of the ground-truth super-resolution image $\bar{\mathbf{f}}$. Remembering that the observed images $\mathbf{g}_n$ are corrupted by mean-zero additive noise $\boldsymbol{\eta}$ we have

$$\begin{aligned} \mathbf{g} &= \Lambda_\alpha \mathtt{M}\bar{\mathbf{f}} + \boldsymbol{\beta} + \boldsymbol{\eta} \\ \mathbf{f}_{\text{mle}} &= (\Lambda_\alpha \mathtt{M})^+ (\mathbf{g} - \boldsymbol{\beta}) \\ &= (\Lambda_\alpha \mathtt{M})^+ (\Lambda_\alpha \mathtt{M}\bar{\mathbf{f}} + \boldsymbol{\beta} + \boldsymbol{\eta} - \boldsymbol{\beta}) \\ &= (\Lambda_\alpha \mathtt{M})^+ (\Lambda_\alpha \mathtt{M}\bar{\mathbf{f}} + \boldsymbol{\eta}) \\ &= \bar{\mathbf{f}} + (\Lambda_\alpha \mathtt{M})^+ \boldsymbol{\eta} \end{aligned} \tag{5.28}$$

Hence the expected estimate is

$$\begin{aligned} \mathrm{E}(\mathbf{f}_{\text{mle}}) &= \mathrm{E}(\bar{\mathbf{f}} + (\Lambda_\alpha \mathtt{M})^+ \boldsymbol{\eta}) \\ &= \mathrm{E}(\bar{\mathbf{f}}) + \mathrm{E}((\Lambda_\alpha \mathtt{M})^+ \boldsymbol{\eta})) \\ &= \bar{\mathbf{f}} + (\Lambda_\alpha \mathtt{M})^+ \mathrm{E}(\boldsymbol{\eta}) \\ &= \bar{\mathbf{f}} \end{aligned} \tag{5.29}$$

The covariance Σ_f of the estimated parameters is obtained as

$$\begin{aligned} \Sigma_f &= \mathrm{Var}[(\Lambda_\alpha \mathtt{M})^+ \boldsymbol{\eta}] \\ &= (\Lambda_\alpha \mathtt{M})^+ \Sigma_\eta (\Lambda_\alpha \mathtt{M})^{+\top} \\ &= (\Lambda_\alpha \mathtt{M})^+ (\sigma^2 \mathtt{I}) (\Lambda_\alpha \mathtt{M})^{+\top} \\ &= \sigma^2 (\Lambda_\alpha \mathtt{M})^+ (\Lambda_\alpha \mathtt{M})^{+\top} \\ &= \sigma^2 (\mathtt{M}^\top \Lambda_\alpha{}^2 \mathtt{M})^{-1} \end{aligned} \tag{5.30}$$

5.10 Predicting the behaviour of the ML estimator

The quality of the super-resolution reconstruction obtained using the ML estimator depends on several factors :

- The noise on the observed images.
- The number of low-resolution images available.
- The zoom ratio S (i.e. the upsampling factor)

- The size and accuracy of σ_{psf}.
- The accuracy of the geometric and photometric registration.

Inaccuracies in the estimation of the PSF and of the registration parameters constitutes *model noise*, and we shall consider these sources of error in Sections 5.11.2 and 5.11.3. In this section, we consider the properties of the ML estimator with respect to the image noise, number of images, zoom ratio and the severity of the blur. The results are verified by appeal to synthetic examples in the next section.

For the sake of clarity we shall omit the photometric parameters in this analysis. Their re-introduction is a trivial exercise for the reader. The normal equations for the maximum likelihood estimator have the form :

$$\mathtt{M}^\top \mathtt{M}\mathbf{f}_{\text{mle}} = \mathtt{M}^\top \mathbf{g}$$

For reasons which will become apparent, we consider the equivalent system

$$\mathtt{K}^{-1}\mathtt{M}^\top \mathtt{M}\mathbf{f}_{\text{mle}} = \mathtt{K}^{-1}\mathtt{M}^\top \mathbf{g}$$

where $\mathtt{K}$ is the diagonal matrix whose non-zero entries are the column sums of $\mathtt{M}$, first introduced in Section 5.7. Hence

$$\mathtt{K}^{-1}\mathtt{M}^\top \mathtt{M}\mathbf{f}_{\text{mle}} = \mathbf{f}_{\text{avg}} \tag{5.31}$$

This allows us to take advantage of the analysis already performed for the average image $\mathbf{f}_{\text{avg}}$. Referring to equation (5.17) we can immediately make the following simplification in the limit of a large number of images :

$$\frac{1}{kN}\mathtt{M}^\top \mathtt{M}\mathbf{f}_{\text{mle}} = \mathbf{f}_{\text{avg}}. \tag{5.32}$$

In the case of zero observation noise, when $\mathbf{f}_{\text{avg}} = \bar{\mathbf{f}}_{\text{avg}}$, it is easily verified that $\mathbf{f}_{\text{mle}} = \bar{\mathbf{f}}$, the ground-truth. In the case of additive observation noise, we can immediately infer that the variance of the reconstruction error $\mathbf{e}_{\text{r}} = \mathbf{f}_{\text{mle}} - \bar{\mathbf{f}}$ will be directly proportional to the variance of the noise on $\mathbf{f}_{\text{avg}}$ by virtue of the linearity of the system:

$$\begin{aligned} \text{Var}(\mathbf{e}_{\text{r}}) &\propto \text{Var}(\mathbf{e}_{\text{avg}}) \\ &= \frac{\sigma^2}{2N\sigma_{\text{psf}}\sqrt{\pi}} \end{aligned}$$

where we have used the result of equation (5.20).

We have seen that $\mathbf{f}_{\text{avg}}$ resembles a blurred version of the required super-resolution image. We would therefore expect that the operator $\frac{1}{kN}\mathtt{M}^\top\mathtt{M}$ resembles the discretizaton of a blur operator, and therefore has an approximate Toeplitz structure [31, 69]. This is indeed the case, as can be seen from figure 5.29, which indicates the sparsity structure of the matrices $\mathtt{M}$ and $\mathtt{M}^\top\mathtt{M}$ for a toy super-resolution problem consisting of a 50×50 super-resolution image, and 10 approximately 25×25 pixel low-resolution images.

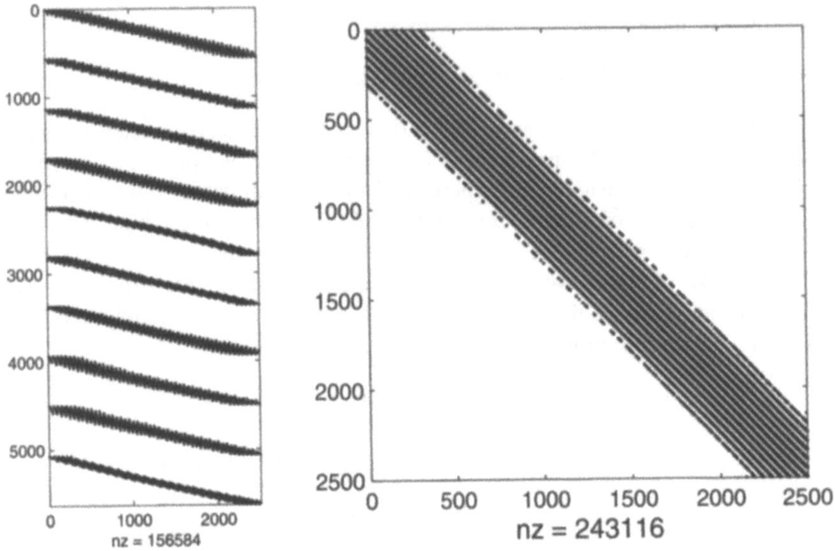

Figure 5.29: The sparsity structure of the matrices $\mathtt{M}$ and $\mathtt{M}^\top\mathtt{M}$ for the toy super-resolution problem described in the text.

To clarify this important point, we have that the ML super-resolution image, pushed through a spatially varying blur operator produces the average image. Hence, the restoration problem is in fact very closely related to the single-image deblurring problem, and the same intuition may be applied.

Given a noisy $\mathbf{f}_{\text{avg}}$, the reconstruction error $\mathbf{e}_{\text{r}}$ will depend of the *condition number* κ of the symmetric, positive-definite matrix $\mathtt{M}^\top\mathtt{M}$. An upper bound on the reconstruction error is given by

$$\frac{||\mathbf{e}_{\text{r}}||}{||\bar{\mathbf{f}}||} \leq \kappa \frac{||\mathbf{e}_{\text{avg}}||}{||\bar{\mathbf{f}}_{\text{avg}}||} \tag{5.33}$$

which is a standard result of linear algebra. A poorly conditioned, near singular matrix (high condition number), will lead to a large reconstruction error, as eigenvectors of $\mathtt{M}^\top\mathtt{M}$ with tiny eigenvalues increasingly corrupt $\mathbf{f}_{\text{mle}}$ as the level of observation noise increases. The character of the reconstruction error will depend on the character of these poorly constrained eigenvectors. Figure 5.30 shows the first 49 eigenvectors of $\mathtt{M}^\top\mathtt{M}$ for the toy 50×50 super-pixel ML estimator. The vectors are ordered by decreasing eigenvalue. Evidently, and as we might have predicted, the eigenvectors (or *eigenimages* as rendered here) resemble a 2D discrete sinusoid basis, with spatial frequency increasing as the eigenvalue decreases. We would therefore expect the reconstruction error to take the form of high-frequency, "checkerboard" patterns.

The condition number κ depends upon σ_{psf}, pixel-zoom ratio S, and the number of images N. The relationships have the following general form :

1. By analogy with single-image deblurring, the greater the severity of the blur (larger σ_{psf}), the more ill-conditioned the inverse problem, and hence the higher

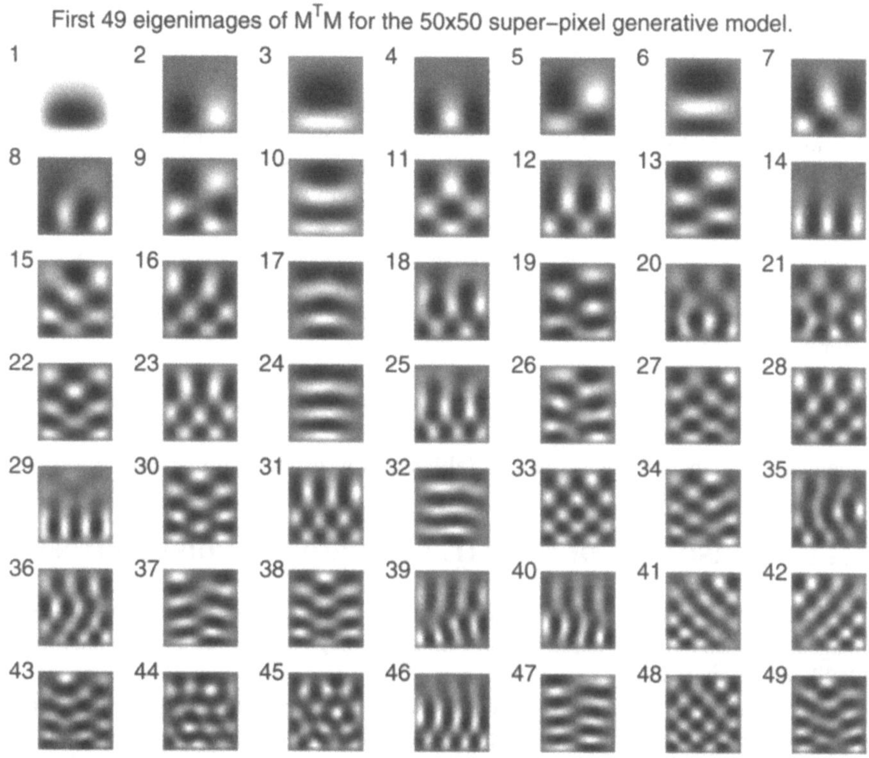

Figure 5.30: The first 49 eigenvectors (eigenimages) of $\mathtt{M}^\top\mathtt{M}$ for the toy 50×50 super-pixel ML estimator. The images resemble a discrete sine or cosine basis, with frequency increasing as the corresponding eigenvalue decreases.

the value of κ.[2]

2. From equation (5.20), we know that the effective noise variance on $\mathbf{e}_{\text{avg}}$ decreases in direct proportion to N, and this in turn will result in a reduced $\mathbf{e}_{\text{r}}$. In tandem with this effect, as the number of images N is increased, the number of equations constraining $\mathbf{f}_{\text{mle}}$ increases roughly in direct proportion to N. Consequently, we would expect κ to decrease as N increases, reaping further reductions in $\mathbf{e}_{\text{r}}$. However, empirical investigation indicates that κ decreases far more slowly than $\frac{1}{N}$ when N is large. Intuitively, the eigenspectrum of $\mathtt{M}^\top\mathtt{M}$ begins to stabilize as more images are added. Hence, in the limit of large N, it is the $\frac{1}{N}$ reduction in $\text{Var}(\mathbf{e}_{\text{avg}})$, on the right-hand side of equation (5.32), that is most important in terms of reducing $\mathbf{e}_{\text{r}}$.

3. As the pixel-zoom ratio S increases, so the *relative size* of σ_{psf}, measured in

[2]This result follows directly from the fact that the eigen-spectrum of a circulant matrix, which encapsulates discrete convolution with some kernel $h(x)$, may be obtained directly as the Fourier transform $H(\omega) \leftrightarrow h(x)$, since any circulant matrix may be diagonalized by a Fourier basis.

terms of super-pixels, also increases, because a single low-resolution pixel covers an increased number of super-pixels. This increases the effective severity of the blur with respect to the super-resolution image, and hence κ increases as explained above. Also, the number of parameters to be estimated increases in proportion to S^2, further increasing κ. The combination of these two effects means that the zoom ratio has by far the largest overall effect on the conditioning of the inverse problem.

To summarize, equations (5.33) and (5.21), together with the relationships explained above, suggest the following form for the upper bound on the reconstruction error $\mathbf{e}_r$:

$$\|e_r\|^2 \leq \frac{c}{S^p N^q \sigma_{\text{psf}}^r} \left(\frac{\sigma}{\pi^{0.25} \sqrt{2N\sigma_{\text{psf}}}} \right) \tag{5.34}$$

where c, p, q, r are unknown constants (although we suspect that $p \geq 2$ and $q << 1$).

5.11 Sensitivity of the ML estimator to noise sources

In this section we examine the effect that various sources of noise, both in the observed images, and also in the parameters controlling the image model, have on the maximum likelihood estimate $\mathbf{f}_{\text{mle}}$.

The maximum likelihood estimate is computed using the *preconditioned conjugate gradient descent* [69] (see Appendix A) algorithm applied to the symmetric, positive-definite system $\mathtt{A}\mathbf{f} = \mathbf{b}$, where $\mathtt{A} = \mathtt{M}^\top \Lambda_\alpha{}^2 \mathtt{M}$ and $\mathbf{b} = \mathtt{M}^\top \Lambda_\alpha (\mathbf{g} - \boldsymbol{\beta})$. The experiments in the following sections employ a simple Jacobi preconditioner, $\text{diag}(\mathtt{A})^{-1}$.

In every case, the algorithm is deemed to have converged when the relative residual falls below 10^{-6}, i.e.

$$\frac{\|\mathtt{A}\mathbf{f} - \mathbf{b}\|_2}{\|\mathbf{b}\|_2} \leq 10^{-6} \tag{5.35}$$

The algorithm terminates when convergence is achieved, or when the iteration count reaches 2000.

5.11.1 Observation noise

The effect of noise in the observed images is investigated by computing the maximum-likelihood estimate from synthetic images which have been degraded with varying levels of additive Gaussian noise. Figure 5.31 shows the ML estimates $\mathbf{f}_{\text{mle}}$ computed using different numbers of images and levels of image noise. The zoom ratio is fixed at 2 in every case. It is clear that the estimator is extremely noise sensitive. Even when using 50 images, a tiny amount of noise severely corrupts the estimate.

From equation (5.30) we expect there to be a linear relationship between the image noise variance and the variance of the ML estimate. This is indeed the case, as demonstrated in Figure 5.32, in which the RMS error between the ground-truth super-resolution image and the ML estimate is plotted against image noise standard-deviation σ.

It is clear from Figures 5.31, 5.32 and 5.33 that the reconstruction error decreases with increased numbers of low-resolution images used. As predicted in Section 5.10, the variance of the reconstruction error decreases in proportion to N when N is large.

Figure 5.34 shows the ML estimates computed using different numbers of images and different pixel zoom ratios. The observation noise is fixed at $\sigma = 0.5$ grey-levels in each case. The reconstruction error decreases rapidly as the zoom ratio is reduced. Figure 5.35 shows the RMS error between the ground-truth and the ML estimate plotted against zoom ratio and number of images. As anticipated in Section 5.10, when the number of images is relatively low (10 to 20), the reconstruction error increases rapidly as the zoom ratio is increased.

5.11.2 Poorly estimated PSF

The point-spread function used in the imaging model can have a pronounced effect on the super-resolution estimate. However, in real imaging situations, the point-spread function is rarely known with any accuracy. Even given an approximate parametric form, such as the Gaussian used here, it can be difficult to estimate the parameter(s) of the PSF from noisy, low-resolution images. The Gaussian PSF adopted in this work has a single parameter σ_{psf} which controls the width of the PSF. In practice, variations in the value of this parameter have a predictable effect on the estimated super-resolution image, as demonstated in Figure 5.36.

The reason for this behaviour is easiest to understand in the case where the input images are related by only Euclidean transformations and the PSF is isotropic. As we have already seen in Section 5.8, in this case, the operations of warping the super-resolution estimate and convolution with the PSF commute. In the case of Euclidean registration there is a family of PSF/super-resolution pairs that can give rise to the same set of observed images. The ML estimator seeks a super-resolution image that, when geometrically warped, filtered by convolution with the PSF, and down-sampled, accurately predicts the observed low-resolution images. If σ_{psf} is too high, the PSF is too wide and hence too "low-pass". In this case, the ML estimate tends to "compensate" by developing excess high-frequency content, which appears as "ringing" artifacts on sharp edges in the super-resolution image. Alternatively, if σ_{psf} is too low, the PSF is too narrow and hence too "high-pass". In this case, the ML estimate does not develop enough high-frequency content, and the super-resolution image is too smooth. A similar effect is observed even when the image-to-image registration is affine or projective, as illustrated by Figure 5.36.

Unfortunately, this ability of the super-resolution ML estimate to compensate for a poorly estimated PSF implies that the log-likelihood score of equation (5.25) tends to be rather insensitive to variations in σ_{psf}. This in turn makes it difficult to include σ_{psf} in the maximum-likelihood estimation. However, since the artifacts induced are quite easy to diagnose by eye, manual tuning of the PSF parameter is straightforward.

5.11.3 Inaccurate registration parameters

To simulate the effect of registration error, zero-mean Gaussian noise, variance σ_H^2 is added to the *translational* components of the synthetic homographies. This at least

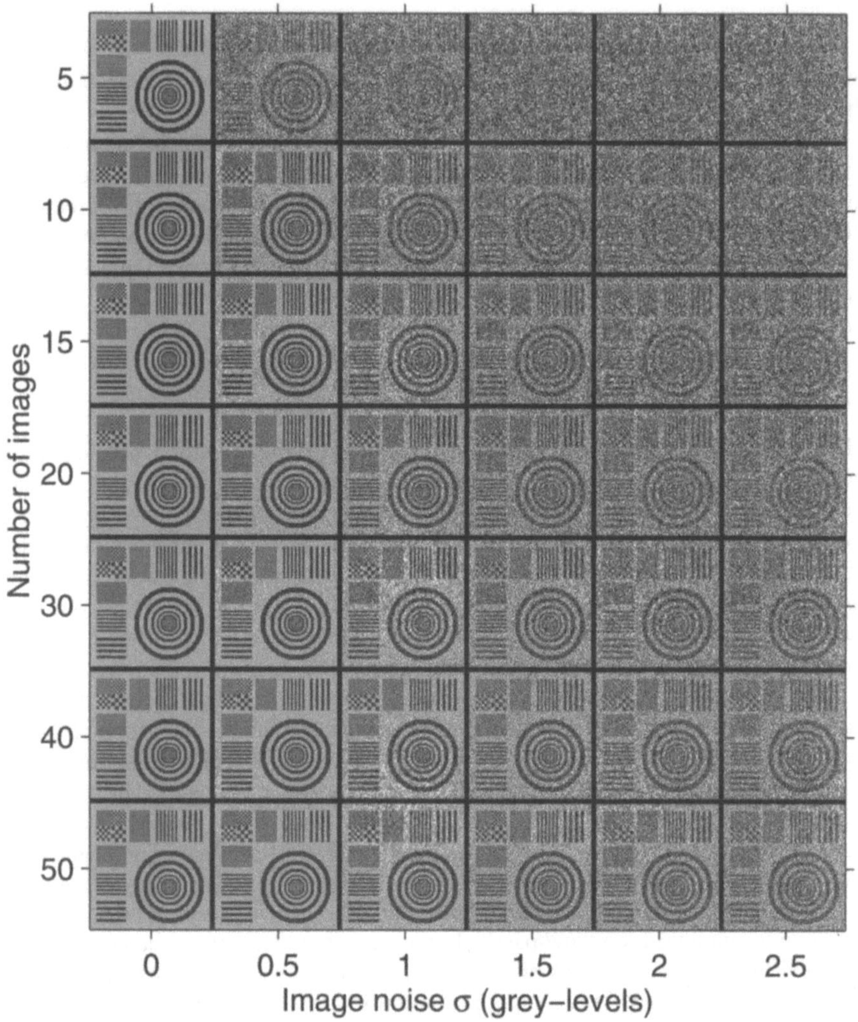

Figure 5.31: A table showing the computed maximum likelihood estimate, $\mathbf{f}_{\mathrm{mle}}$ given different combinations of image noise standard deviation σ, and number of low-resolution images used N. The zoom ratio S was fixed at 2. The results demonstrate the extreme sensitivity of the estimator to noise in the observed images.

gives σ_H some intuitive meaning. Figure 5.37 shows the ML estimate computed from noise-free 50 synthetic images for increasing levels of registration error. The zoom ratio is fixed at 1.5. As expected, the reconstruction error increases with increased σ_H, but unlike the error induced by observation noise, it is not homogeneously distributed. This is due to the fact that in smooth regions of the image, errors in registration make very little difference in the simulated pixel values. Only in the vicinity of edges does the registration error have a significant effect on the simulated pixels, and it is around

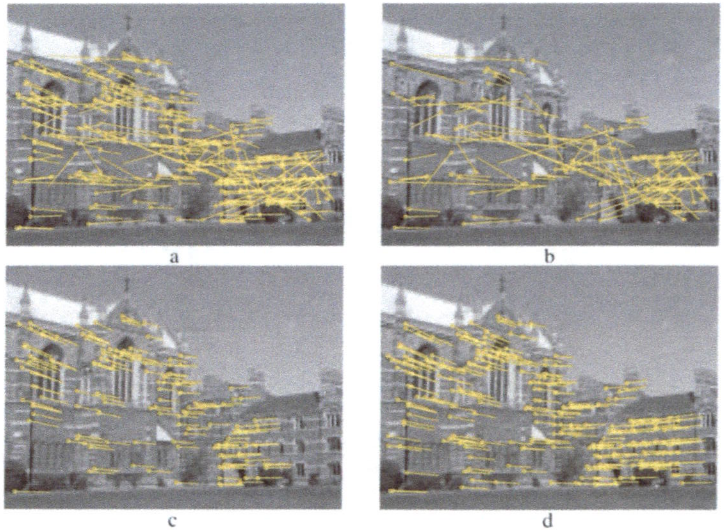

Plate 1: (a) A winner-takes-all scheme is applied to the ranked disparity feature matches. The 268 putative matches are shown superimposed on the left image by the line-linking matched points. Note the clear mismatches. (b) RANSAC outliers: 117 of the putative matches. (c) RANSAC inliers: 151 correspondences consistent with the estimated $\mathtt{H}$. (d) The final set of 262 correspondences after guided matching and MLE. The estimated $\mathtt{H}$ is accurate to sub-pixel resolution.

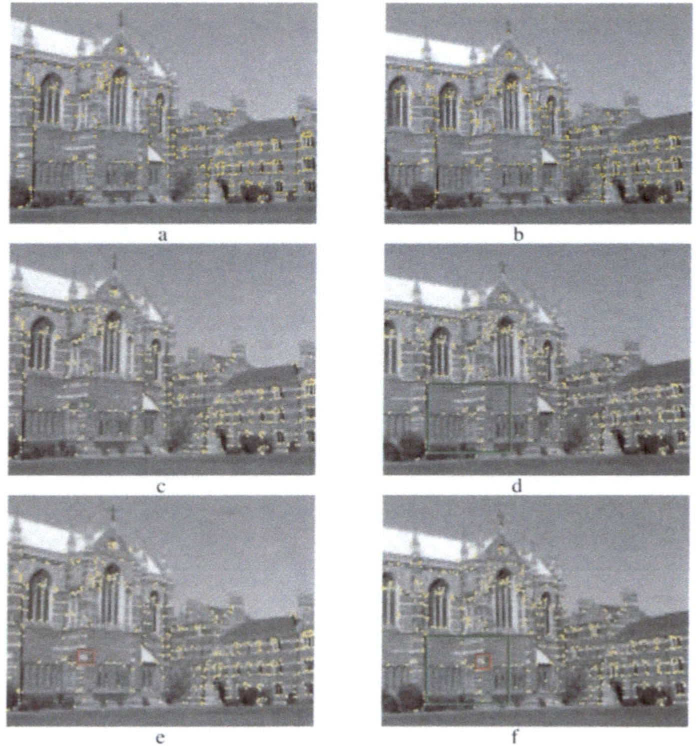

Plate 2: (a,b) Images of Keble College, Oxford. The motion between views is a rotation about the camera centre so the images are related by a homography. The images are 640×480 pixels. Detected Harris interest points superimposed on the images. There are approximately 500 features on each image. (c) An interest point in the left image, and (d) the corresponding search window in the right image (shown in green). (e,f) Possible matches are ranked by computing the normalized cross-correlation between 5×5 pixel windows centred on the two interest points (shown in red).

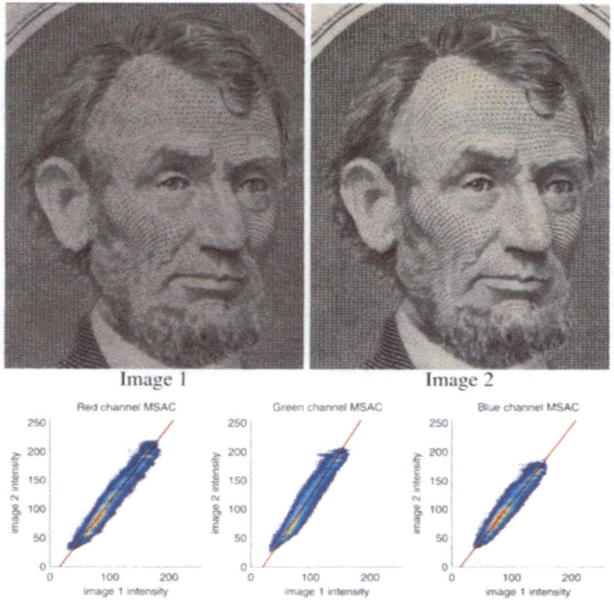

Plate 3: A pair of images of two physically different bank-notes. The image portrayed is the same, but one has been treated with a dye, causing significant photometric difference. (Top) The geometrically registered images. (Bottom) MSAC line fit to intensities (i_1, i_2) at corresponding pixel locations in the two images. The colour indicates the density of observations of corresponding intensities (i_1, i_2).

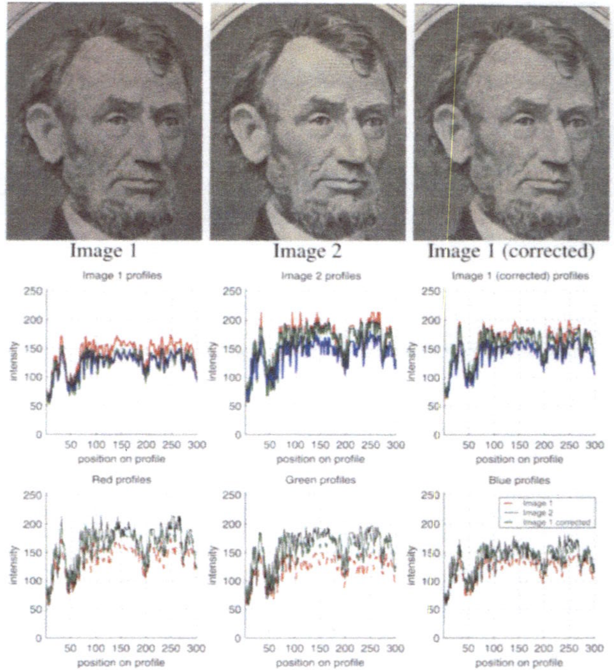

Plate 4: (Top) The original and corrected versions of image 1, either side of image 2. The corrected version has lost its pink hue, closely resembling image 2, and just revealing a purple fingerprint mark on the face. (Middle) Intensity-profile plots along a slice (shown above in yellow) through each of the images. The profiles in the corrected image resemble far more closely those of image 2 than those in the original image. (Bottom) Intensity-profiles from the original image (red) and corrected image (green) overlaid on the profiles from image 2 (black). The corrected profiles are a very good fit.

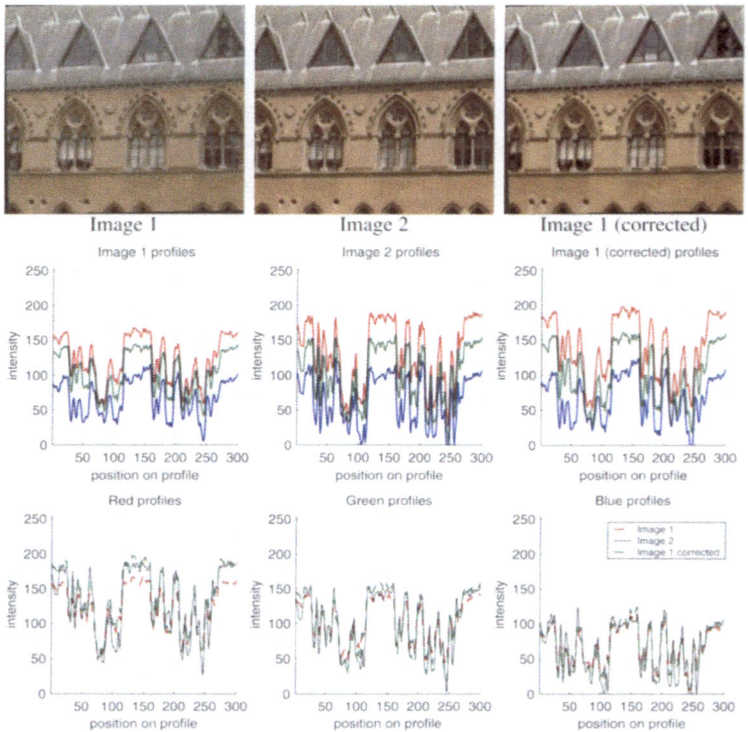

Plate 5: A pair of images with significant photometric differences due to a change in camera white-balance. (Top) The original and corrected versions of image 1, either side of image 2. The corrected version has lost its undesirable yellow hue, closely resembling image 2. (Middle) Intensity-profile plots along a slice (shown above in yellow) through each of the images. The profiles in the corrected image resemble far more closely those of image 2 than those in the original image. (Bottom) Intensity-profiles from the original image (red) and corrected image (green) overlaid on the profiles from image 2 (black). The corrected profiles are a very good fit.

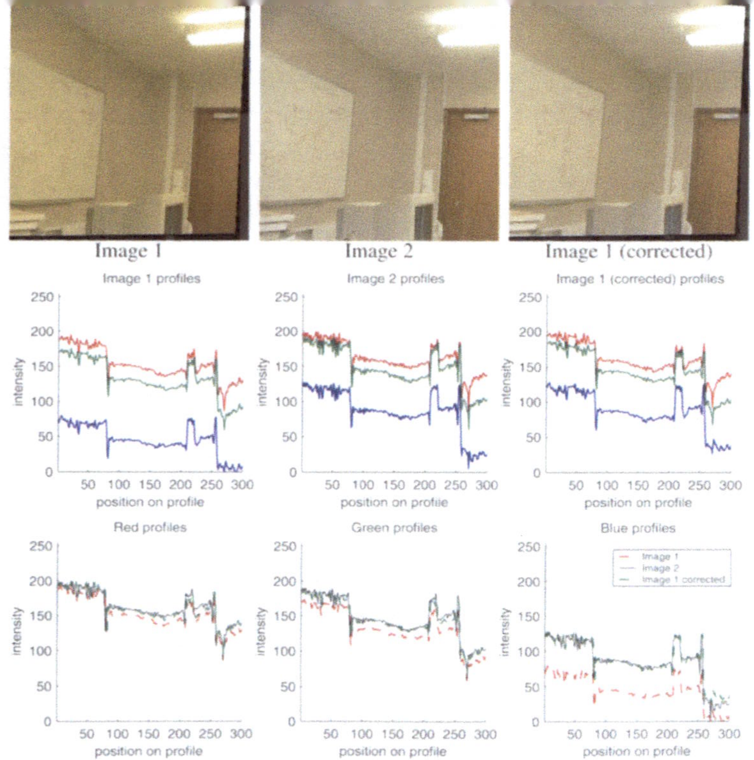

Plate 6: A pair of images with significant photometric differences due to a change in illumination conditions. (Top) The original and corrected versions of image 1, either side of image 2. The corrected version has the orange glow of the sun-lit image 2. (Middle) Intensity-profile plots along a slice (shown above in yellow) through each of the images. The profiles in the corrected image resemble far more closely those of image 2 than those in the original image. (Bottom) Intensity-profiles from the original image (red) and corrected image (green) overlaid on the profiles from image 2 (black). The corrected profiles are a very good fit.

(a) evidence (b) control

(c) registered control (d) difference image

Plate 7: (a) A British five-pound note image captured using a flat-bed scanner. The latent finger prints have been enhanced with a special dye, but are still difficult to interpret due to the confusing background. (b) The control image is a clean note captured in the same manner. (c) The control image after geometric and photometric registration to align it with the evidence. (d) The difference image quite clearly shows the latent finger mark. In this example, the relative displacements of the text layer and the dominant background patterns are slightly different in the two notes. Consequently, the text layer is not registered by the computed homography, and the text is still visible in the difference image. This is an artifact of the bank note printing process.

Plate 8: Six primary colour images blended using three common techniques: (Top) nearest image-centre, (Middle) simple averaging, (Bottom) feathering.

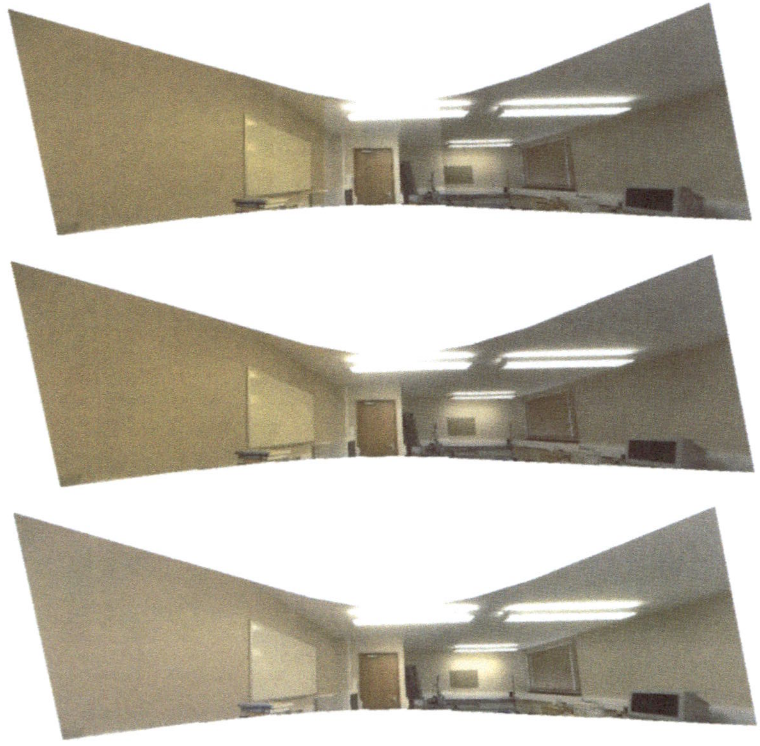

Plate 9: (Top) A mosaic rendered from eight images using the nearest image-centre blending method. Seams due to photometric differences are clearly visible. (Middle) The mosaic rendered using feathering. The problem is ameliorated, but the unsightly variation of colour balance across the image still remains. (Bottom) The same mosaic rendered using photometrically corrected images. Frame 5 was used as the photometric reference frame. No seams are visible and the colour balance is uniform across the image.

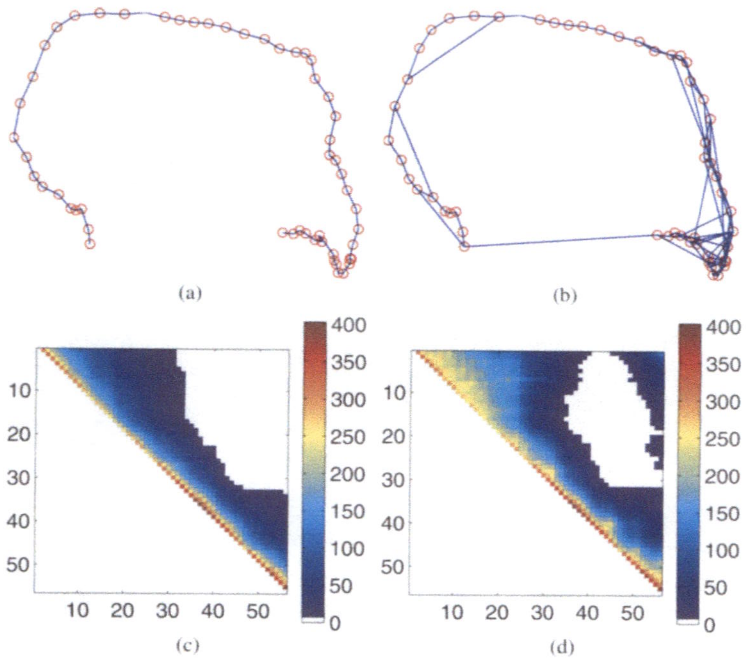

Plate 10: Results of applying the *N*-view matching algorithm to the "Oxford Map" sequence. (a) and (b) show the view graph before and after application of the *N*-view matching algorithm. The image centres are represented by the circles. The explicitly computed matches populate the graph with edges. The initial view graph contains only edges between consecutive image pairs. The described *N*-view matching algorithm inserts an additional 45 edges into the graph by performing matching between image pairs where it is considered to be advantageous according to the criterion described in Equation (4.6). (c) and (d) represent the number of feature matches between every pair of frames as a colour-coded matrix.

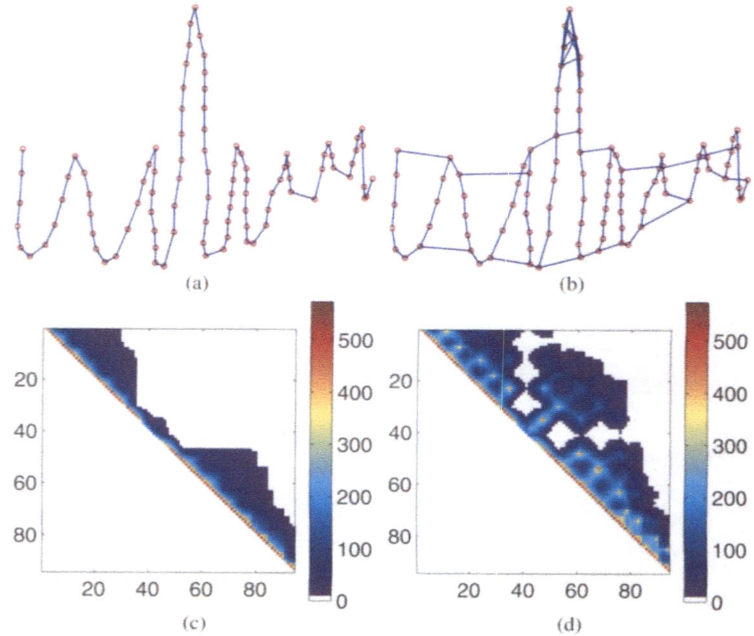

Plate 11: (a) and (b) show the view graphs for the University Museum sequence before and after *N*-view matching. The algorithm inserts an additional 25 edges into the view-graph. (c) and (d) represent the number of feature matches between every pair of frames as a colour-coded matrix. The amount of off-diagonal mass is significantly increased, indicating many extra matches between non-consecutive image pairs.

Plate 12: (Top) The mosaic formed by composing the 94 images using temporal median filtering. (Bottom) The outline of every 5th frame is overlaid.

Plate 13: A planar manifold mosaic composed from 44 images.

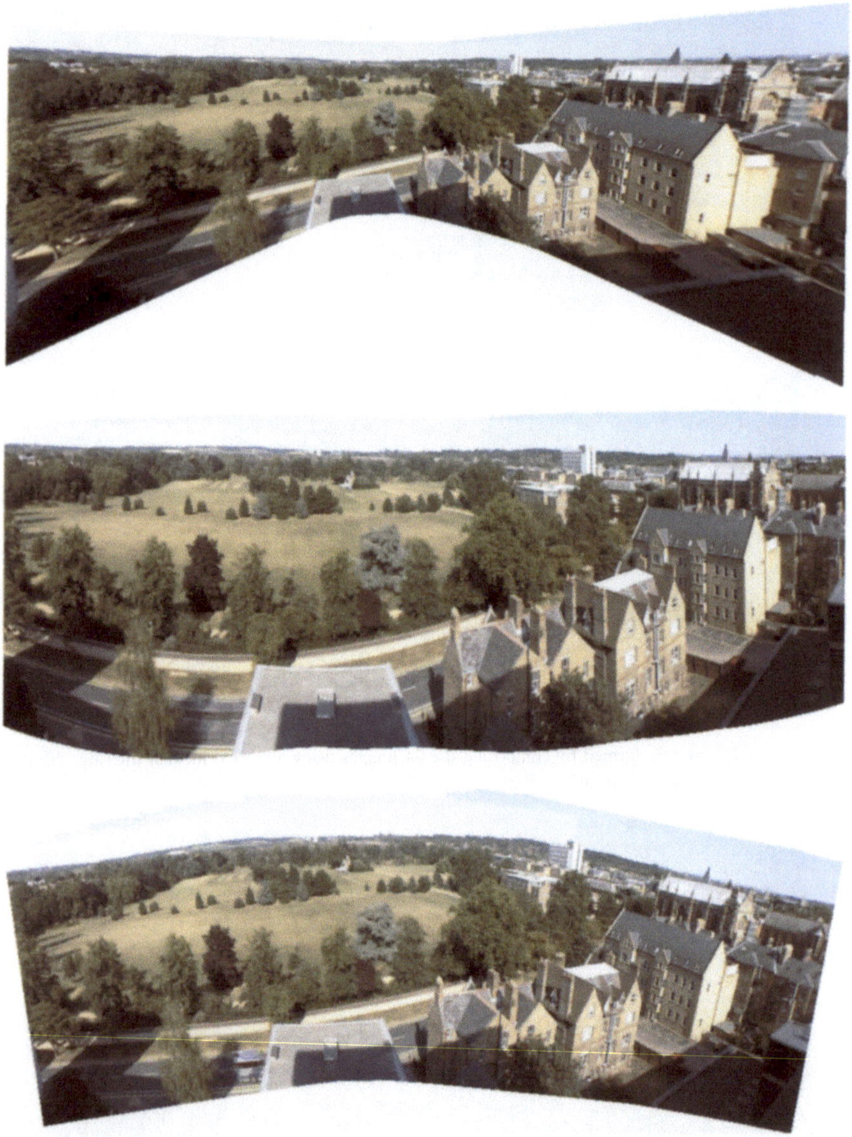

Plate 14: Mosaics created from a sequence of 43 images captured using a hand-held MiniDV camera. (Top) Planar manifold projection. (Middle) Cylindrical manifold. (Bottom) Peleg's manifold projection method.

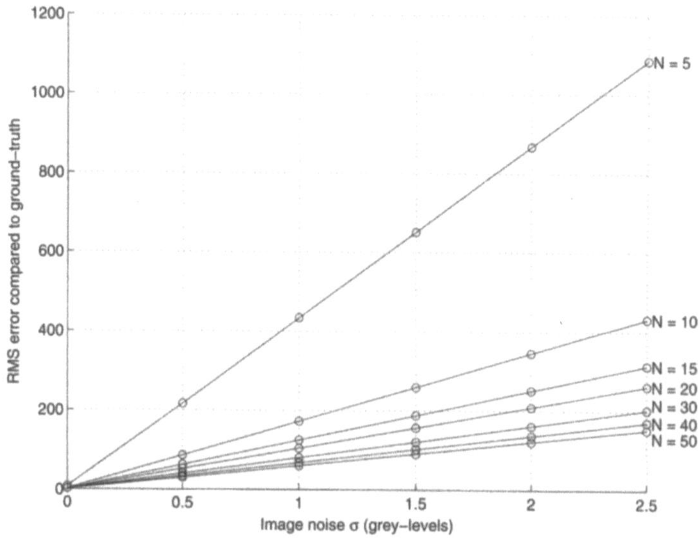

Figure 5.32: The RMS of the ground-truth error ($\mathbf{f}_{\mathrm{mle}} - \bar{\mathbf{f}}$) is plotted against the additive Gaussian noise standard-deviation σ in the low-resolution images for several different values of N, the number of images used. The zoom ratio S was fixed at 2. As one might expect given the linearity of the system, there is a perfect linear relationship between the reconstruction error and the observation noise.

the edges that the reconstruction error tends to be most severe. The graph shows the variation of ground-truth error with respect to σ_H. Interestingly, the error increases slowly for at first, becoming fairly linear for larger values of σ_H. Note that, even when $\sigma_H = 0$, the reconstruction error is not zero. This is due to the bicubic interpolation method which is used to compare the reconstructed 96×96 with the original 128×128.

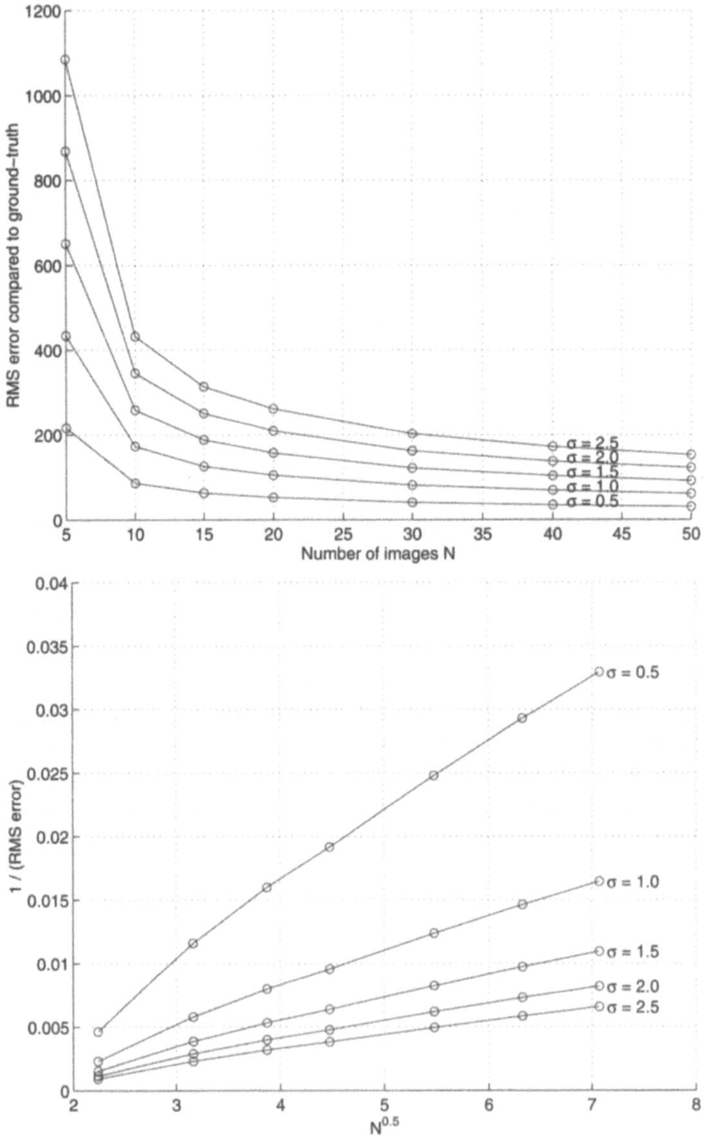

Figure 5.33: (Top) The RMS of the ground-truth error $(\mathbf{f}_{\text{mle}} - \bar{\mathbf{f}})$ is plotted against the number of images used N, for a range of image noise standard-deviations σ. The zoom ratio S was fixed at 2. (Bottom) When plotted as the reciprocal of RMS error against $\sqrt{N}$ it is clear that, for $N > 10$, the reconstruction error is inversely proportional to the square-root of the number of images used.

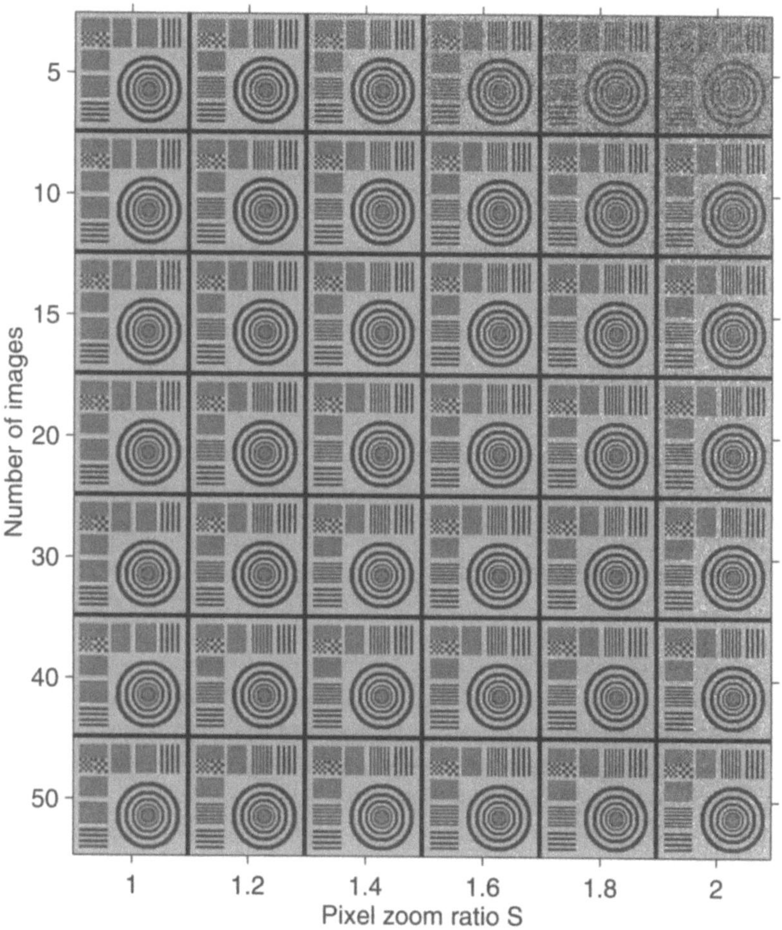

Figure 5.34: A table showing the ML estimate given different combinations of pixel zoom ratio S and number of images N. The image noise was fixed at $\sigma = 0.5$ grey-levels. The reconstruction error decreases rapidly as the zoom ratio is reduced.

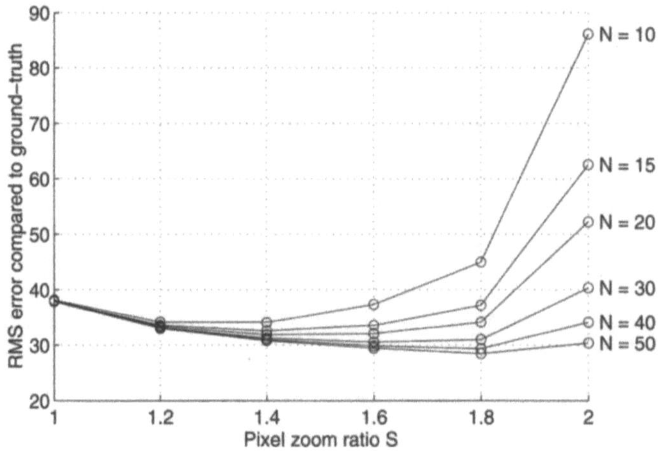

Figure 5.35: The RMS error between the ML estimate and the ground-truth is plotted against zoom ratio and number of images. When the number of images is relatively small (10 to 20) the error increases rapidly as the zoom ratio increases.

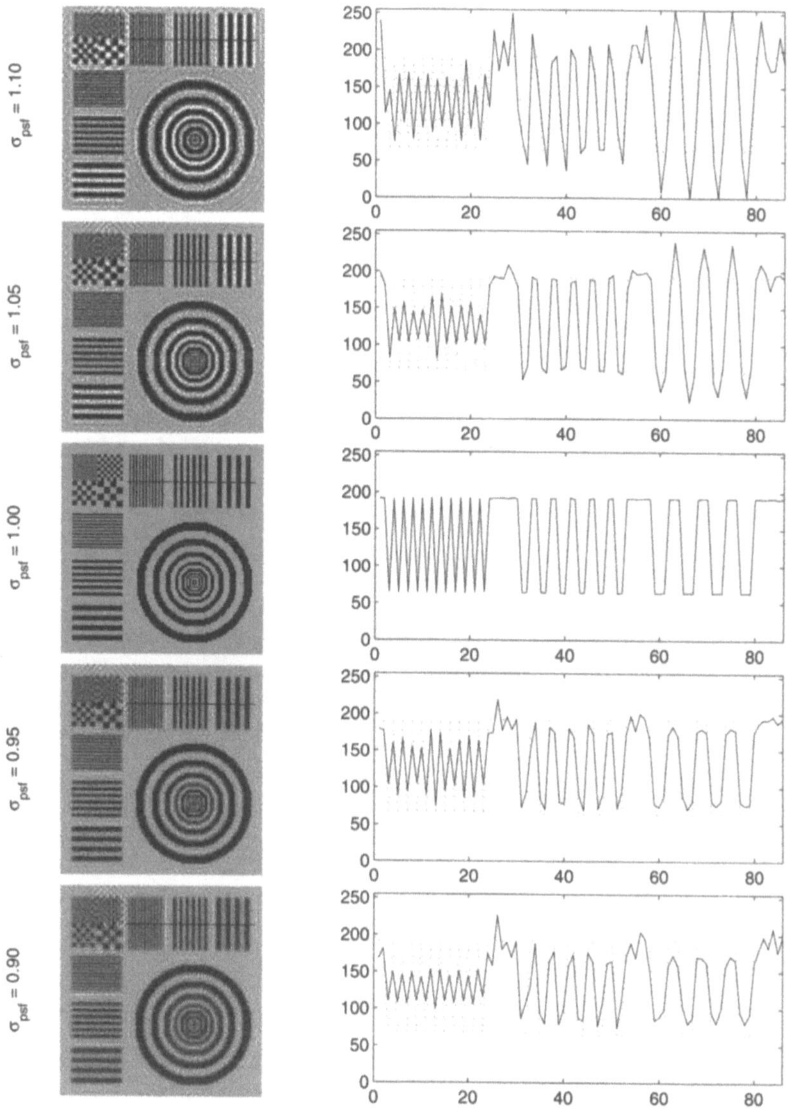

Figure 5.36: The effect of a poorly estimated point-spread function is investigated by computing the maximum likelihood estimate for several different values of σ_{psf}, the size of the Gaussian PSF. The zoom ratio S was fixed at 2, and 50 noise-free images were used. The correct σ_{psf} is 1.0. When the PSF is too wide (too "low-pass"), the estimate exhibits high-frequency "ringing" artifacts.When the PSF is too narrow (too "high-pass"), the estimate is blurred. These effects are clarified in the intensity-profile plots on the right. The dotted profile is the ground-truth.

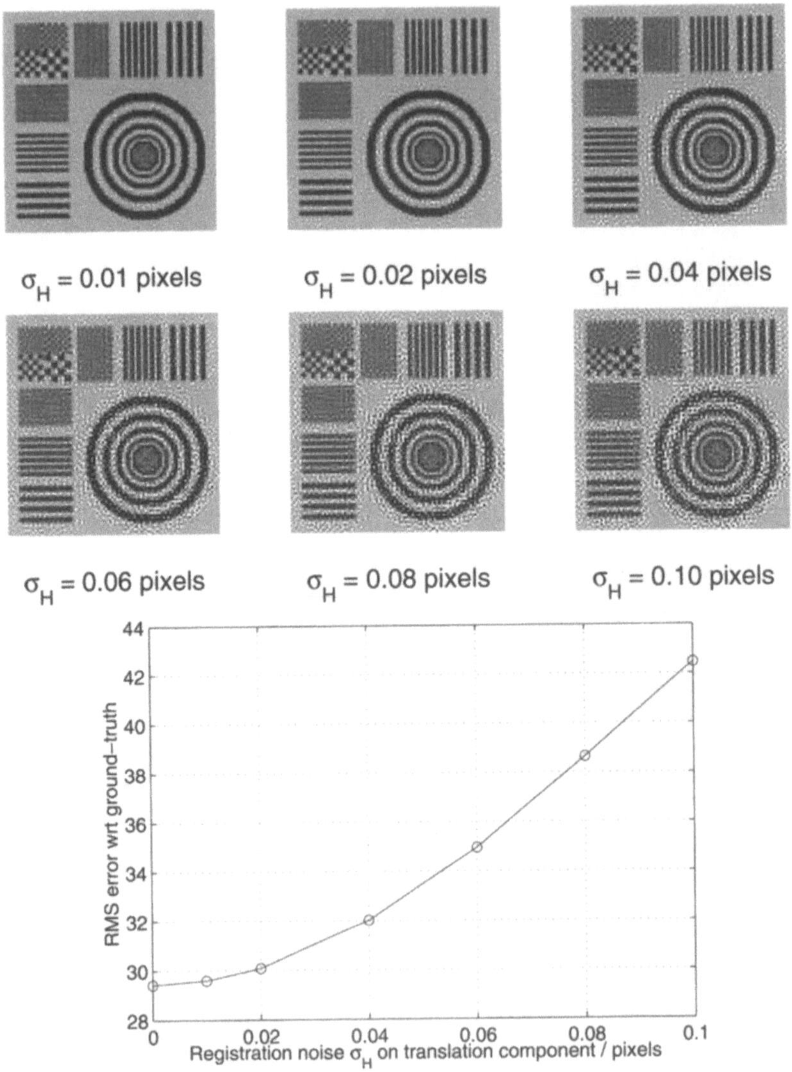

Figure 5.37: (Top) The ML estimate using 50 noise-free synthetic images at $1.5\times$ zoom for increasing levels of registration error σ_H. The reconstruction error is not homogeneous, being more severe around edges in the image. (Bottom) The RMS error with respect to the ground-truth is plotted against σ_H. The reconstruction error increases slowly at first, becoming fairly linear for larger σ_H.

5.12 Irani and Peleg's method

In this section we review the extremely popular method proposed by Irani and Peleg [84, 85]. We show that, under certain circumstances, their method minimizes the same cost function as the ML estimator (although the illumination parameters are omitted), and also examine the convergence properties. For brevity, we refer to their methods as the IP algorithm.

In the IP algorithm, the iterative update of the super-resolution estimate proceeds by an error back-projection scheme inspired by computer-aided tomography. When all the low-resolution images have been simulated, the residual images (simulated minus observed) are convolved with a back-projection function (BPF) and warped back into the super-resolution frame. The back-projected errors from all the observed images are averaged and used to directly update the estimate as follows

$$\mathrm{s}^{i+1} = \mathrm{s}^i + \frac{1}{C}\sum_{\forall n} \mathcal{T}_n^{-1}[h_{\mathrm{bpf}} * \mathrm{S}\!\uparrow\!(\hat{\mathrm{m}}_n - \mathrm{m}_n)] \tag{5.36}$$

where C is a constant and h_{bpf} is the back-projection kernel. Irani and Peleg suggest that $h_{\mathrm{bpf}} = (h_{\mathrm{psf}})^k$ where $k \geq 1$ is a good choice of BPF, ensuring convergence whilst suppressing spurious noise components in the solution.

Before embarking on an analysis of Irani and Peleg's estimator, we briefly review the method of least-squares minimization by steepest descent, which we later refer to in the analysis.

5.12.1 Least-squares minimization by steepest descent

An over-constrained least-squares problem has the form

$$\hat{\mathbf{x}} = \arg\min_{\mathbf{x}} C(\mathbf{x}) = \|\mathbf{b} - \mathtt{A}\mathbf{x}\|^2$$

The solution to this problem is easily obtained by differentiating the cost function $C(\mathbf{x})$ with respect to $\mathbf{x}$:

$$\frac{\mathrm{d}C}{\mathrm{d}\mathbf{x}} = -2\mathtt{A}^\top(\mathbf{b} - \mathtt{A}\mathbf{x}) = 0$$
$$\Rightarrow \hat{\mathbf{x}} = (\mathtt{A}^\top\mathtt{A})^{-1}\mathtt{A}^\top\mathbf{b}$$

When the system is very large, the solution cannot be obtained by direct means. A simple, iterative procedure for finding the vector $\hat{\mathbf{x}}$ which is the solution to this minimization problem is by the method of *steepest descent*. The direction of steepest descent on the error surface C is

$$\mathbf{d} = \mathtt{A}^\top(\mathbf{b} - \mathtt{A}\mathbf{x}) \tag{5.37}$$

The algorithm proceeds by taking successive small steps in the direction of steepest descent :

$$\mathbf{x}^{i+1} = \mathbf{x}^i + k\mathtt{A}^\top(\mathbf{b} - \mathtt{A}\mathbf{x}^i) \tag{5.38}$$

where the constant $k < 1$ is chosen appropriately in order to ensure convergence.

Fixed point Setting $\mathbf{x}^{i+1} = \mathbf{x}^i$ gives a fixed point $\mathbf{x}^\infty$

$$\begin{aligned} \mathbf{x}^\infty &= \mathbf{x}^\infty + k\mathtt{A}^\top(\mathbf{b} - \mathtt{A}\mathbf{x}^\infty) \\ &= (\mathtt{A}^\top\mathtt{A})^{-1}\mathtt{A}^\top\mathbf{b} \end{aligned} \tag{5.39}$$

Hence the fixed point corresponds to the least-squares solution.

Convergence Convergence is analysed with respect to the fixed point solution $\mathbf{x}^\infty$

$$\begin{aligned} \mathbf{e}^i &= \mathbf{x}^\infty - \mathbf{x}^i \\ \mathbf{x}^i &= \mathbf{x}^\infty - \mathbf{e}^i \\ \mathbf{x}^\infty - \mathbf{e}^{i+1} &= \mathbf{x}^\infty - \mathbf{e}^i + k\mathtt{A}^\top(\mathbf{b} - \mathtt{A}(\mathbf{x}^\infty - \mathbf{e}^i)) \\ \mathbf{e}^{i+1} &= \mathbf{e}^i - k\mathtt{A}^\top\mathtt{A}\mathbf{e}^i \\ &= (\mathtt{I} - k\mathtt{A}^\top\mathtt{A})\mathbf{e}^i \end{aligned} \tag{5.40}$$

Hence

$$\|\mathtt{I} - k\mathtt{A}^\top\mathtt{A}\| < 1 \tag{5.41}$$

is a condition for convergence. This is equivalent to saying that the largest eigenvalue (spectral radius) of $\mathtt{I} - k\mathtt{A}^\top\mathtt{A}$ must be inside the unit circle.

Decomposing the matrix $\mathtt{A}$ in terms of its SVD, and projecting the error vectors $\mathbf{e}^i$ onto the right-eigenspace gives

$$\mathtt{A} = \mathtt{U\Sigma V}^\top \tag{5.42}$$

$$\mathbf{e}^i = \sum_j e^i_j \mathbf{v}_j \tag{5.43}$$

The convergence scheme may then be rewritten as

$$\begin{aligned} \sum_j e^{i+1}_j \mathbf{v}_j &= (\mathtt{I} - k\mathtt{V\Sigma}^\top\mathtt{U}^\top\mathtt{U\Sigma V}^\top)\sum_j e^i_j\mathbf{v}_j \\ &= (\mathtt{I} - k\sum_j \sigma_j^2 \mathbf{v}_j\mathbf{v}_j^\top)\sum_j e^i_j\mathbf{v}_j \end{aligned}$$

which, by orthogonality, yields

$$e^{i+1}_j = (1 - k\sigma_j^2)e^i_j \tag{5.44}$$

Hence the eigenvectors with the largest corresponding singular values converge fastest.

5.12.2 Irani and Peleg's algorithm

The IP algorithm proceeds as follows

$$\begin{aligned} f^{i+1} &= f^i + k\frac{\sum_n \mathcal{T}_n^{-1} b * \; s\!\uparrow(g_n - \hat{g}_n)}{\sum_n \mathcal{T}_n^{-1} b * \; s\!\uparrow(1)} \\ &= f^i + k\frac{\sum_n \mathcal{T}_n^{-1} b * \; s\!\uparrow(g_n - \; s\!\downarrow(h * \mathcal{T}_n f^i))}{\sum_n \mathcal{T}_n^{-1} b * \; s\!\uparrow\downarrow} \end{aligned} \tag{5.45}$$

where b is the back-projection function. The function ${}_s\!\uparrow\downarrow$ is the *0-1 comb-function,* a uniform grid of delta functions of height 1, with zeros everywhere else. The denominator $\sum_N \mathcal{T}_n^{-1} b * {}_s\!\uparrow\downarrow$ amounts to a set of normalizing constants which may be precalculated. In matrix form the algorithm becomes

$$\mathbf{f}^{i+1} = \mathbf{f}^i + k\mathtt{B}^\top(\mathbf{g} - \mathtt{M}\mathbf{f}^i) \tag{5.46}$$

where $\mathtt{B}^\top$ is the back-projection matrix, combining the up-sampling, convolution and inverse geometric transformation. $\mathtt{B}^\top$ is row-normalized so that each row is performing a weighted average of a set of error-residuals.

Fixed point The fixed point of the algorithm is given by

$$\begin{aligned} \mathbf{f}^\infty &= \mathbf{f}^\infty + k\mathtt{B}^\top(\mathbf{g} - \mathtt{M}\mathbf{f}^\infty) \\ &= (\mathtt{B}^\top\mathtt{M})^{-1}\mathtt{B}^\top\mathbf{g} \end{aligned} \tag{5.47}$$

Under our zero-mean, additive noise assumptions, we can subtitute $\mathbf{g} = \mathtt{M}\bar{\mathbf{f}} + \boldsymbol{\eta}$ and assuming a choice of $\mathtt{B}$ such that $\mathtt{B}^\top\mathtt{M}$ is full-rank, we obtain

$$\begin{aligned} \mathbf{f}^\infty &= (\mathtt{B}^\top\mathtt{M})^{-1}\mathtt{B}^\top(\mathtt{M}\bar{\mathbf{f}} + \boldsymbol{\eta}) \\ &= \bar{\mathbf{f}} + (\mathtt{B}^\top\mathtt{M})^{-1}\mathtt{B}^\top\boldsymbol{\eta} \end{aligned} \tag{5.48}$$

where $\bar{\mathbf{f}}$ is the ground-truth super-resolution image. The expectation of the fixed point is then

$$\begin{aligned} \mathrm{E}(\mathbf{f}^\infty) &= \mathrm{E}(\bar{\mathbf{f}} + (\mathtt{B}^\top\mathtt{M})^{-1}\mathtt{B}^\top\boldsymbol{\eta}) \\ &= \bar{\mathbf{f}} + (\mathtt{B}^\top\mathtt{M})^{-1}\mathtt{B}^\top\mathrm{E}(\boldsymbol{\eta}) \\ &= \bar{\mathbf{f}} \end{aligned}$$

Hence, the algorithm is an unbiased estimator of the ground-truth.

Convergence properties Again, the convergence of the algorithm is analysed with respect to the fixed point solution

$$\begin{aligned} \mathbf{e}^i &= \mathbf{f}^\infty - \mathbf{f}^i \\ \mathbf{f}^i &= \mathbf{f}^\infty - \mathbf{e}^i \\ \mathbf{f}^\infty - \mathbf{e}^{i+1} &= \mathbf{f}^\infty - \mathbf{e}^i + k\mathtt{B}^\top(\mathbf{g} - \mathtt{M}(\mathbf{f}^\infty - \mathbf{e}^i)) \\ \mathbf{e}^{i+1} &= \mathbf{e}^i - k\mathtt{B}^\top\mathtt{M}\mathbf{e}^i \\ &= (\mathtt{I} - k\mathtt{B}^\top\mathtt{M})\mathbf{e}^i \end{aligned} \tag{5.49}$$

Hence

$$||\mathtt{I} - k\mathtt{B}^\top\mathtt{M}|| < 1 \tag{5.50}$$

is a condition for convergence. Clearly, if $\mathtt{B}^\top = \mathtt{M}^\top$ the IP algorithm is the same as steepest descent (5.41).

5.12.3 Relationship to the ML estimator

In the case when $\mathtt{B}^\top = \mathtt{M}^\top$, we can see by comparing equation (5.48) with equation (5.28) (neglecting the unused photometric parameters Λ_α), that the IP algorithm gives exactly the same solution as the ML estimator.

The effect of the back-projection operator $\mathtt{B}^\top$ may be misconstrued as having some regularizing effect on the solution. To a certain extent this is true. In theory, one could choose a $\mathtt{B}^\top$ whose column-span does not include certain troublesome vectors (say, very high-frequency components). It would then be impossible for these components to appear in $\mathbf{f}$. However, it is hard to imagine exactly how such a $\mathtt{B}^\top$ might be chosen.

In practice then, $\mathtt{B}^\top$ will be full-rank. As seen in equation (5.48), $\mathtt{B}^\top$ has no effect on the expected $\mathbf{f}$ over repeated trials. It does however affect the character of the reconstruction error $\mathbf{e}_{\text{ip}} = (\mathtt{B}^\top\mathtt{M})^{-1}\mathtt{B}^\top\boldsymbol{\eta}$. One way to think about the effect of $\mathtt{B}^\top$ is as follows. Imagine that the noise $\boldsymbol{\eta}$ on our low-resolution pixels is not independent and identically distributed, but rather has some covariance Σ_η. In this case, the ML estimate would be obtained by minimizing

$$\mathcal{L} = (\mathbf{g} - \mathtt{M}\mathbf{f})^\top \Sigma_\eta (\mathbf{g} - \mathtt{M}\mathbf{f})$$

for which the stationary point occurs at

$$\frac{\mathrm{d}\mathcal{L}}{\mathrm{d}\mathbf{f}} = \mathtt{M}^\top\Sigma_\eta\mathtt{M}\mathbf{f} - \mathtt{M}^\top\Sigma_\eta\mathbf{g} = 0$$

Substituting $\mathtt{B}^\top = \mathtt{M}^\top\Sigma_\eta$, we obtain

$$\begin{aligned} \mathtt{B}^\top(\mathbf{g} - \mathtt{M}\mathbf{f}) &= 0 \\ \Rightarrow \mathbf{f} &= (\mathtt{B}^\top\mathtt{M})^{-1}\mathtt{B}^\top\mathbf{g} \end{aligned}$$

which is identical to the fixed point of the IP algorithm in equation (5.47). To summarize then, the IP algorithm can always be thought of as an ML estimator, but one in which the choice of $\mathtt{B}^\top$ implies some underlying assumption about the noise covariance Σ_η of the observed low-resolution pixels, namely that $\mathtt{B}^\top = \mathtt{M}^\top\Sigma_\eta$. For the choice of back-projection function suggested in the authors' original paper, $h_{\text{bpf}} = h^2_{\text{psf}}$, $\mathtt{B} \neq \mathtt{M}$ and the algorithm does not perform ML estimation under independent noise assumptions.

5.12.4 Convergence properties

In the case when $\mathtt{B} = \mathtt{M}$, the only difference between the ML estimator and the IP algorithm is that the latter implies a particular method of reaching the optimum: fixed step-length, gradient descent. Consequently, it is *several orders of magnitude* slower to converge than the conjugate gradient algorithm applied to the same problem. This fact is demonstrated with a synthetic example using 20 low-resolution images, corrupted by additive noise $\sigma = 0.5$ grey-levels. The reconstruction is performed at zoom ratio $S = 2$. Figure 5.38 shows the ground-truth RMS error plotted against number of iterations for the ML estimator using conjugate gradient descent, and the

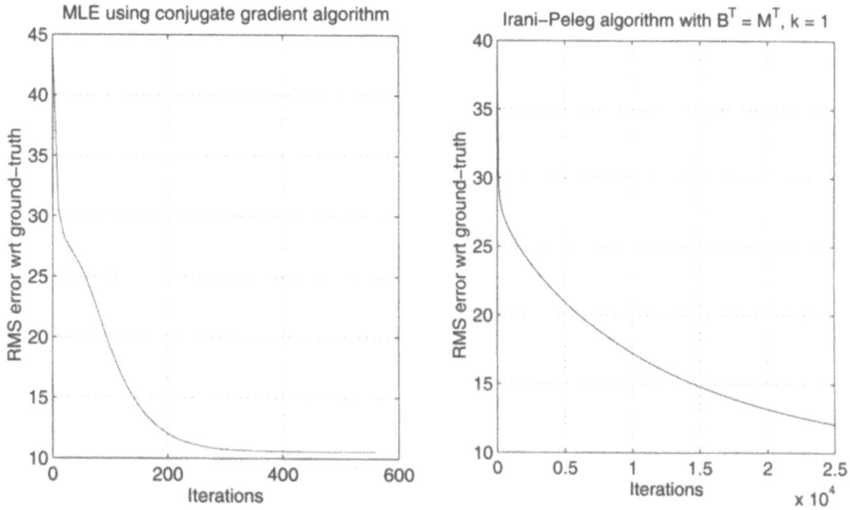

Figure 5.38: Comparison of the rates of convergence of the ML estimator using conjugate gradient descent, and the IP algorithm, which is the same as fixed step-length steepest descent. Note the difference in scale on the horizontal axes. Even after 25 000 iterations, the IP algorithm has yet to converge.

IP algorithm with $\mathtt{B}^\top = \mathtt{M}^\top$. The value of k was manually tuned to the largest value for which the algorithm still converges. Note the difference in scale of the horizontal axes. Conjugate gradients converges within 600 iterations. By comparison, even after 25 000 iterations, the IP algorithm has yet to converge.

5.13 Gallery of results

Figures 5.39 to 5.46 show a variety of results obtained using the ML super-resolution estimator applied to four different real image sequences. These results demonstrate that, despite its shortcomings, it is possible to obtain reconstructions using the MLE which show marked improvement in detail over the original low-resolution images, even though the levels of pixel zoom are quite modest.

The median image In some of the examples, the super-resolution results are compared with the "median image", obtained by geometrically warping/resampling the input images into the super-resolution coordinate frame, and combining them using a median filter.

Note on the presentation of low-resolution images Throughout this section, and indeed the rest of this book , in order to give a fair representation of the performance of the super-resolution algorithms when compared with a "text book" image-zooming method, low-resolution input images are shown *up-sampled* using bicubic interpolation to the resolution of the corresponding super-resolution results.

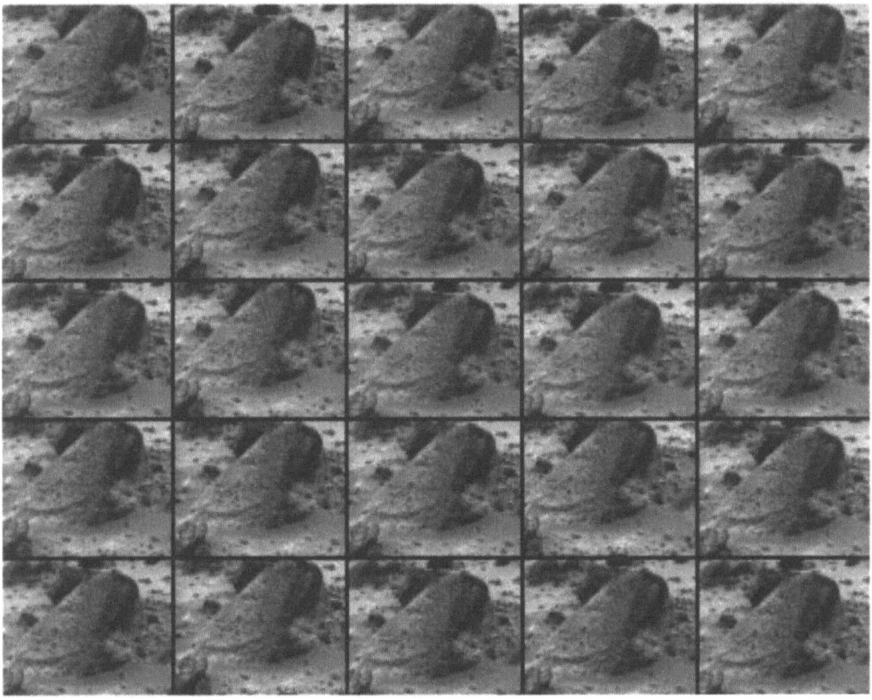

Figure 5.39: A sequence of 25 images captured by the NASA Mars Pathfinder (available in the NSSDC image catalogue). The images show the "Wedge" rock captured by successive sweeps of a rotating camera.

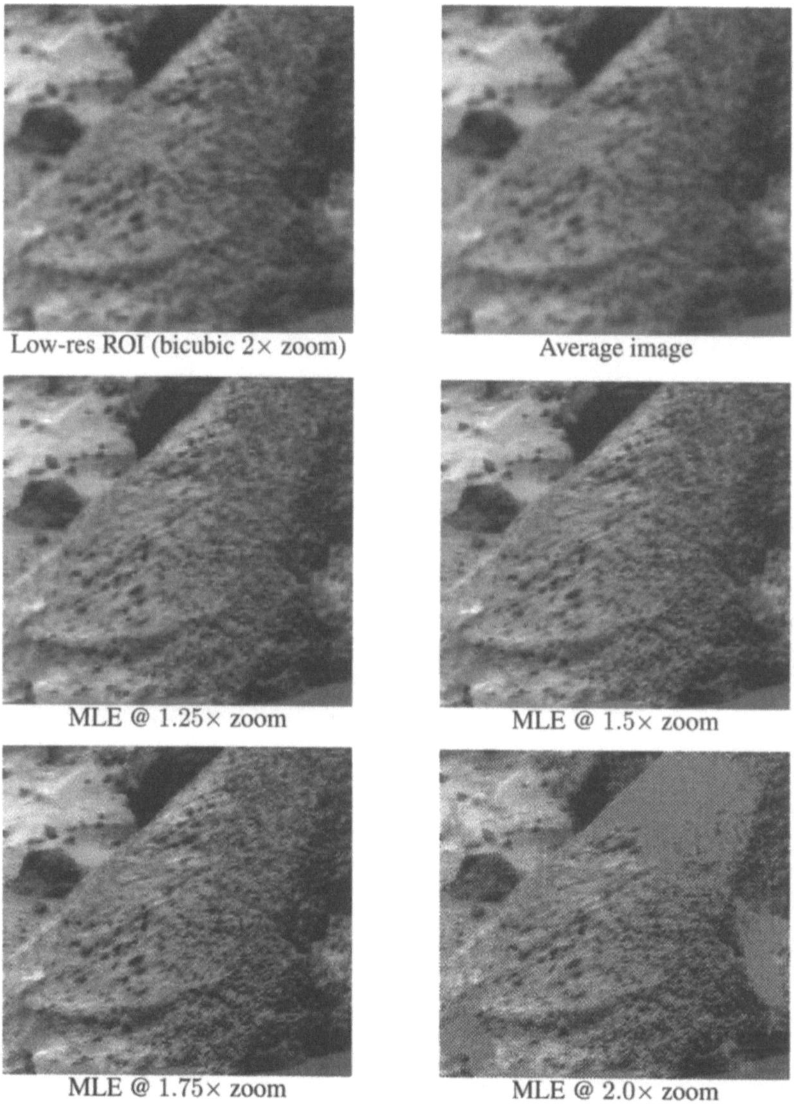

Figure 5.40: The ML estimate obtain from the "Wedge" sequence, reconstructed at four different pixel zoom levels. The results up to 1.75× zoom show a marked improvement in detail over the low-resolution image and average image. The result at 2× zoom is badly corrupted by "checkerboard" reconstruction error. σ_{psf} was set to 0.45 for this sequence. For comparison the low-resolution region-of-interest (ROI) is shown bicubically up-sampled to 2× zoom.

Figure 5.41: (Top) 16 frames from a sequence of 200 captured at the University of Surrey, using a hand-held digital video camera. (Bottom) A mosaic composed from the entire sequence. The region-of-interest (boxed) contains a car.

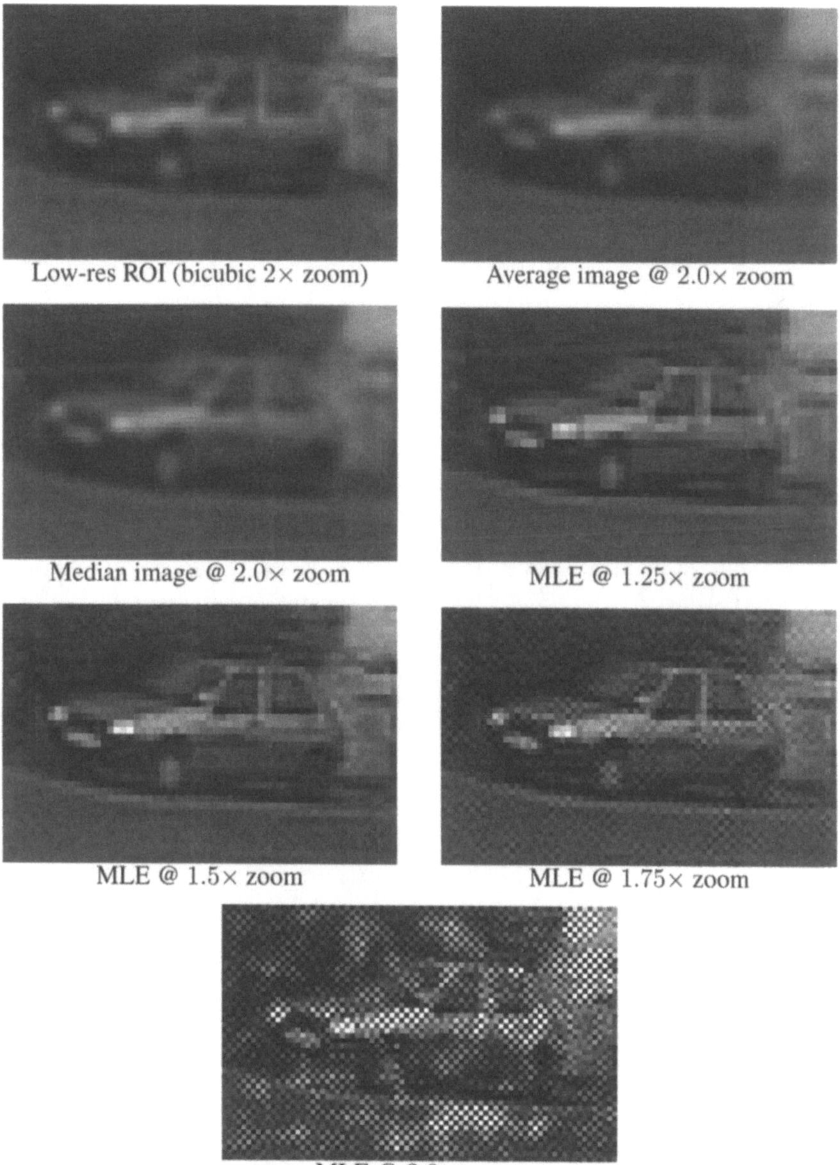

Figure 5.42: ML super-resolution estimates computed using 50 low-resolution images. Reconstructions up to $1.5\times$ zoom show marked improvement over the low-resolution, average and median images. Reconstruction error starts to become apparent at $1.75\times$ zoom. σ_{psf} was set to 0.425 for this sequence.

Figure 5.43: (Top) 16 frames from a sequence of 250 captured at the University of Surrey, using a hand-held digital video camera. (Bottom) A mosaic composed from the entire sequence. The region-of-interest (boxed) contains some illegible text.

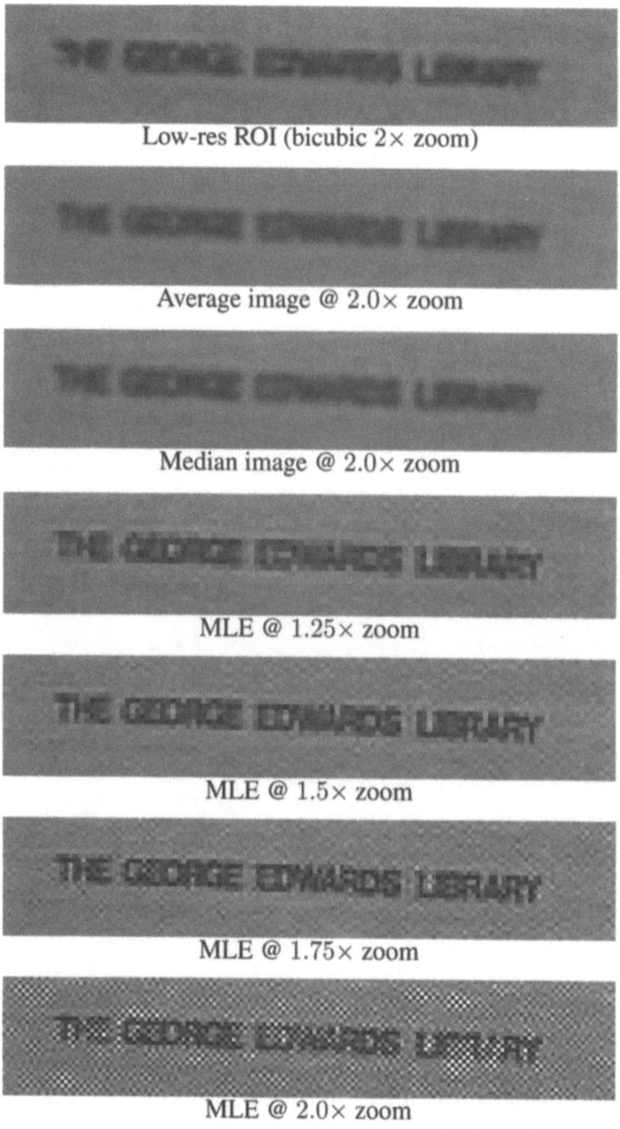

Figure 5.44: ML super-resolution estimates computed using 30 low-resolution images. Reconstructions up to 1.75× zoom show marked improvement over the low-resolution, average and median images. Reconstruction error starts to become apparent at 2.0× zoom. σ_{psf} was set to 0.4 for this sequence.

Figure 5.45: (Top) One of a sequence of 10 images captured by Tomas Pajdla and Daniel Martinec at CMP, Prague. The camera is fitted with a Nikon 360° lens. The rig is designed such that the CCD array can be micro-translated independently of the lens system. (Bottom) A close-up view of the test-card.

Low-res ROI (bicubic 2× zoom)

Average image @ 2.0× zoom

Median image @ 2.0× zoom

MLE @ 1.25× zoom

MLE @ 1.5× zoom

Figure 5.46: ML reconstructions using 10 images. The small number of input images means that reconstruction error becomes apparent at quite modest zoom. σ_{psf} was set to 0.4 for this sequence. The median image produces a marginally sharper result than the average image in this example.

5.14 Summary

In this chapter we have examined a maximum-likelihood estimator for super-resolution reconstruction, based on a generative model of the imaging process. The implementation issues regarding the generative model have been discussed, and an accurate and efficient implementation proposed. The sensitivity of the ML estimate to noise on the observed images, and errors in the model (registration and point-spread function) has been analysed and empirically demonstrated using a variety of synthetic examples. The classic super-resolution algorithm of Irani and Peleg has also been analysed, and the conditions under which it converges to the ML estimate have been explained. The algorithm was demonstrated to have an extremely slow rate of convergence compared with the conjugate gradient scheme which is used throughout this book . Finally, the ML estimator was applied to several real image sequences, and was shown to produce reasonable results provided the pixel zoom ratio is relatively low.

In the next chapter, we investigate the use of a Bayesian framework for super-resolution restoration, and see how statistical image models may be used to dramatically improve the condition of the problem.

Chapter 6

Super-resolution Using Bayesian Priors

6.1 Introduction

This chapter introduces the use of Bayesian prior image models in the super-resolution reconstruction problem. As seen in Chapter 5, the maximum-likelihood estimator of the super-resolution image is a highly ill-conditioned inverse problem. Consequently, the solution is extremely sensitive to noise in the observed images and to errors in registration. It is notable that, in cases where the ML estimator performs poorly, the resulting super-resolution image does not resemble what we would normally consider a "sensible" image. It is this observation that motivates the Bayesian super-resolution methods described in this chapter. Whereas the ML estimator only considers the conditional likelihood of the observations with respect to the super-resolution image, the Bayesian methods introduce an additional probability density defined *a priori* over the space of all images. The combined distribution over possible super-resolution images is termed the *posterior*, and the *maximum a posterior* (MAP) estimator provides a solution which is both a good fit to the observations, and also has a high likelihood with respect to the prior image model. The image priors considered in this chapter are simple Markov random field (MRF) models, based on heuristic notions about the short-range spatial correlations between pixels in "typical" images.

In Section 6.2, we show how the introduction of a Bayesian prior modifies the super-resolution reconstruction problem. In Section 6.3, we expose the relationship between certain MAP estimators and the classical method of optimal (Weiner) filtering, demonstrating how this method can be re-cast in a Bayesian framework. In Section 6.4, we examine some common image priors which have been proposed in both the image restoration and super-resolution literature, and demonstrate the particular applicability of the Huber MRF model [29, 131] for the super-resolution reconstruction of images containing text. In Section 6.6 we demonstrate the improved noise robustness which is obtained using the MAP estimators. In Section 6.7, we describe a novel algorithm which uses cross-validation to automatically set the "hyper-parameter" which balances the influence of the prior model against the influence that the observed images have on the reconstructed image. Finally, Section 6.8 demonstrates the MAP estimators applied to several real image sequences.

6.2 The Bayesian framework

In this section we derive the general form of the maximum *a posterior* estimator for the super-resolution image. From equation (5.24), the total probability of an observed image $g_n(x,y)$ given an estimate of the super-resolution image $\hat{f}(x,y)$ is determined by the assumed Gaussian distribution of the image noise

$$\Pr(\mathbf{g}_n|\hat{\mathbf{f}}) = \prod_{\forall x,y} \frac{1}{\sigma\sqrt{2\pi}} \exp\left(-\frac{(\hat{g}_n(x,y) - g_n(x,y))^2}{2\sigma_n^2}\right)$$

where the simulated low-resolution image $\hat{\mathbf{g}}_n$ is given by

$$\hat{\mathbf{g}}_n = \mathtt{M}_n\hat{\mathbf{f}}$$

Note that, throughout this and the subsequent chapter, we shall drop the explicit photometric parameters, (α_n, β_n), in order to improve the clarity of the equations presented. Putting them back in is a trivial exercise for the reader. Of course, the algorithms used to generate the results in this chapter do still include the photometric parameters in their computations, and in the real examples they are estimated robustly using the method described in Chapter 3.

Assuming independent observations, the probability over all images in the sequence is

$$\Pr(\mathbf{g}|\hat{\mathbf{f}}) = \prod_{\forall n} \Pr(\mathbf{g}_n|\hat{\mathbf{f}}) \tag{6.1}$$

where $\mathbf{g}$ is the stack of vectors $\mathbf{g}_n$. The prior image model provides an additional pdf defined over the space of all images, $\Pr(\mathbf{f})$. By invoking Bayes's theorem [91, 94, 118, 138], we can combine the two to obtain the posterior distribution :

$$\Pr(\hat{\mathbf{f}}|\mathbf{g}) = \frac{\Pr(\mathbf{g}|\hat{\mathbf{f}})\Pr(\hat{\mathbf{f}})}{\Pr(\mathbf{g})} \tag{6.2}$$

The *maximum a posterior* (MAP) estimate of $\mathbf{f}$ is then

$$\begin{aligned} \mathbf{f}_{\text{map}} &= \arg\max_{\mathbf{f}} \frac{\Pr(\mathbf{g}|\hat{\mathbf{f}})\Pr(\hat{\mathbf{f}})}{\Pr(\mathbf{g})} \\ &= \arg\max_{\mathbf{f}} \Pr(\mathbf{g}|\hat{\mathbf{f}})\Pr(\hat{\mathbf{f}}) \end{aligned} \tag{6.3}$$

Taking logarithms we obtain

$$\begin{aligned} \mathbf{f}_{\text{map}} &= \arg\max_{\mathbf{f}} \quad \lg\Pr(\hat{\mathbf{f}}) + \lg\Pr(\mathbf{g}|\hat{\mathbf{f}}) \\ &= \arg\max_{\mathbf{f}} \quad \lg\Pr(\hat{\mathbf{f}}) - \frac{1}{2\sigma_n^2}\|\mathtt{M}\mathbf{f} - \mathbf{g}\|^2 \end{aligned} \tag{6.4}$$

The specific form of $\lg \Pr(\hat{\mathbf{f}})$ depends on the prior being used. All the priors in this chapter, and in fact the majority of those in the image restoration literature, are Markov random field priors.

6.2.1 Markov random fields

Our basic image model is a random field, a collection of independent random variables. The Markov random field concept tightens this with the assumption that the conditional pdf of a single pixel, conditioned on *all* other pixels in the image, is equal to the pdf conditioned on just some *subset* of the other pixels, i.e.

$$\Pr(f_i | f_{j \neq i}) = \Pr(f_i | f_k, k \in \mathcal{N}_i) \tag{6.5}$$

The neighbourhood $\mathcal{N}_i$ is termed the *Markov blanket* of pixel f_i. In some sense, MRFs are the N-dimensional analogue of the familiar 1-d Markov chain. In MRF image models, the pixel neighbourhoods $\mathcal{N}$ are typically determined by some spatial adjacency rule. They are invariably spatially homogeneous, meaning that every pixel has the same neighbourhood structure regardless of its position in the image. Common examples are the 4-neighbour and 8-neighbour blankets.

This formulation of the MRF in terms of a conditional pdf is rather inconvenient when it comes to actually defining a prior which has particular properties, such as smoothness, or piecewise constancy. For this reason, most MRF models are expressed in terms of the equivalent *Gibbs distribution*.

6.2.2 Gibbs priors

Any MRF can be expressed as an equivalent Gibbs distribution[1] which has the form

$$\Pr(\mathbf{f}) = \frac{1}{Z} \exp\left(- \sum_{\forall C \in \mathcal{C}} V_C(C) \right) \tag{6.6}$$

where Z is a normalizing constant (called the *partition function* in statistical mechanics), C are *cliques* of images pixels, the $V_C(C)$ are *potential functions* defined over the pixels in each clique, and $\mathcal{C}$ is the set of all cliques contained within the image $\mathbf{f}$. A clique is set of pixels which are mutual neighbours according to some adjacency rule. Common image priors are defined using only pair-cliques, meaning that each V is a functions of only two pixel values, $V(C) = V(f_i, f_j)$. They are typically *homogeneous*, meaning that the $V(C)$ depend only upon the clique type and not upon the position of the clique in the image. They are typically *isotropic*, meaning that the $V(C)$ depend only on the distance between the pair of pixels in C and not upon its orientation. For further details on the modelling of images with Gibbs distributions the reader is referred to Besag [14, 15], Geman and Geman [65] and Li [96].

In the case when the prior is a Gibbs distribution, equation (6.4) becomes

$$\mathbf{f}_{\text{map}} = \arg\max_{\mathbf{f}} \; - \sum_{\forall C \in \mathcal{C}} V_C(C) - \frac{1}{2\sigma_n^2} \|\mathtt{M}\mathbf{f} - \mathbf{g}\|^2 \tag{6.7}$$

[1] The equivalence of MRFs and Gibbs distributions is known as the *Hammersley–Clifford theorem* [96].

The difference between the many image priors proposed in the literature essentially comes down to the size and spatial extent of the cliques used, and the nature of the potential functions. The latter fall into three categories: *quadratic, non-quadratic convex* and *non-convex*. In the following sections we shall see examples of each of these three cases.

6.2.3 Some common cases

The simplest and most common Gibbs priors have potential functions that are quadratic in the pixel values $\mathbf{f}$, hence

$$\Pr(\mathbf{f}) = \frac{1}{Z} \exp\left(-\mathbf{f}^{\top} \mathtt{Q} \mathbf{f}\right) \tag{6.8}$$

where $\mathtt{Q}$ is a symmetric, positive-definite matrix. In this case, equation (6.7) becomes

$$\mathbf{f}_{\text{map}} = \arg\max_{\mathbf{f}} \quad -\hat{\mathbf{f}}^{\top} \mathtt{Q} \hat{\mathbf{f}} - \frac{1}{2\sigma_n^2} ||\mathtt{M}\mathbf{f} - \mathbf{g}||^2 \tag{6.9}$$

This case is of particular interest since the MAP estimator has, in principle, a linear solution:

$$\mathbf{f}_{\text{map}} = \left(\mathtt{M}^{\top}\mathtt{M} + \mathtt{Q}\right)^{-1} \mathtt{M}^{\top} \mathbf{g} \tag{6.10}$$

Of course, in the context of image restoration, it is computationally infeasible to perform the matrix inversion directly, but since both terms in equation (6.9) are quadratic, the conjugate gradient ascent method may be applied to obtain the solution iteratively.

The simplest matrix $\mathtt{Q}$ which satisfies the criterion is a multiple of the identity, giving

$$\mathbf{f}_{\text{map}} = \arg\max_{\mathbf{f}} \quad -\gamma^2 ||\mathbf{f}||^2 - \frac{1}{2\sigma_n^2} ||\mathtt{M}\mathbf{f} - \mathbf{g}||^2 \tag{6.11}$$

A common variation on this scheme is when $\mathtt{Q}$ is derived from a linear operator $\mathtt{L}$ applied to the image $\mathbf{f}$:

$$\mathbf{f}_{\text{map}} = \arg\max_{\mathbf{f}} \quad -\gamma^2 ||\mathtt{L}\mathbf{f}||^2 - \frac{1}{2\sigma_n^2} ||\mathtt{M}\mathbf{f} - \mathbf{g}||^2 \tag{6.12}$$

in which case $\mathtt{Q}$ is $\mathtt{L}^{\top}\mathtt{L}$. The matrix $\mathtt{L}$ is typically chosen to be a discrete approximation of a first or second derivative operator. Equations (6.11) and (6.12) will be familiar to many people as forms of *Tikhonov regularization* [57, 71, 149] , a technique proposed by Tikhonov and Arsenin in the context of solving Fredholm integral equations of the first kind. Image deconvolution is one example of this class of problem.

Another way to think about equation (6.8) is as a multi-variate Gaussian distribution over $\mathbf{f}$, in which $\mathtt{Q}$ is the inverse of the covariance matrix.

6.3 The optimal Wiener filter as a MAP estimator

In this section we briefly review the Wiener filtering method, a classical frequency domain technique which is used to "regularize" the ill-conditioned image deconvolution problem. We show how this method can be cast as MAP estimation with a Gibbs prior, and conversely, how certain Gibbs priors may be interpreted as priors in the frequency domain.

The Wiener filter or "optimal least-squares filter" [70, 88, 115] has proved to be a very successful tool in the reconstruction of single blurred, noisy images. It is a frequency domain method which provides an estimate of the deblurred image which minimizes the error between the reconstruction and the ground-truth in a least-squares sense. The method requires accurate knowledge of the point-spread function, and approximate knowledge of the power spectral densities of the ground-truth image and the additive noise corrupting the observed image. As with other frequency domain restoration methods, the blurring is limited to a linear convolution with a point-spread function and must therefore be spatially invariant.

The degraded image g is modelled as

$$g = h * f + n \tag{6.13}$$

where f is the undegraded image, h is the point-spread function, n is an additive noise term and $*$ is the convolution operator. Transforming the problem to the frequency domain, we have

$$G(\omega) = H(\omega)F(\omega) + N(\omega) \tag{6.14}$$

The Wiener method finds a reconstruction filter $Q(\omega)$, which is applied to the degraded image $G(\omega)$ to obtain the reconstructed image $\hat{F}(\omega)$:

$$\begin{aligned}\hat{F}(\omega) &= Q(\omega)G(\omega) \\ &= Q(\omega)(H(\omega)F(\omega) + N(\omega))\end{aligned} \tag{6.15}$$

The expected reconstruction error is

$$\begin{aligned}r &= \mathbb{E}[\|\hat{F}(\omega) - F(\omega)\|^2] \\ &= \mathbb{E}[\|Q(\omega)(H(\omega)F(\omega) + N(\omega)) - F(\omega)\|^2]\end{aligned} \tag{6.16}$$

where $\mathbb{E}[.]$ is the expectation operator. It is easily shown (see [70]) that the optimal choice of $Q(\omega)$, which minimizes the reconstruction error in a least-squares sense, is

$$Q(\omega) = \frac{H(\omega)^* P_F(\omega)}{H(\omega)^* P_F(\omega) H(\omega) + P_N(\omega)} \tag{6.17}$$

where H^* indicates the complex conjugate, and $P_N(\omega)$ and $P_F(\omega)$ are the power spectral densities (PSD)[2] of the noise and signal respectively.

Given this choice of $Q(\omega)$ the reconstructed image is

$$\hat{F}(\omega) = \frac{H(\omega)^* P_F(\omega) G(\omega)}{H(\omega)^* P_F(\omega) H(\omega) + P_N(\omega)} \tag{6.18}$$

[2]The power spectral density of a signal $F(\omega)$ is the expectation of its auto-correlation spectrum, $P_F(\omega) = \mathbb{E}[F(\omega)^* F(\omega)]$.

It is straightforward to show that $\hat{F}(\omega)$ is the solution of the following frequency domain restoration problem :

$$\hat{F}(\omega) = \arg\min_{F} \left\| \frac{H(\omega)F(\omega) - G(\omega)}{\sqrt{P_N(\omega)}} \right\|^2 + \left\| \frac{F(\omega)}{\sqrt{P_F(\omega)}} \right\|^2 \tag{6.19}$$

Using Parseval's theorem, and remembering that component-wise multiplication (or division) in the Fourier domain is equivalent to convolution (or deconvolution) in the spatial domain, we can rewrite this equation as an equivalent spatial domain image restoration problem:

$$\hat{\mathbf{f}} = \arg\min_{\mathbf{f}} \quad (\mathtt{H}\mathbf{f} - \mathbf{g})^\top \Lambda_n^{-1} (\mathtt{H}\mathbf{f} - \mathbf{g}) + \mathbf{f}^\top \Lambda_f^{-1} \mathbf{f} \tag{6.20}$$

in which Λ_n^{-1} and Λ_f^{-1} are the circulant convolution matrices whose rows are the inverse Fourier transforms of $P_N(\omega)$ and $P_F(\omega)$. We can now see that the PSD prior imposed in the Wiener filter corresponds to an image noise covariance matrix Λ_n and a Gibbs prior of the form

$$\Pr(\mathbf{f}) = \frac{1}{Z} \exp(-\mathbf{f}^\top \Lambda_f^{-1} \mathbf{f}) \tag{6.21}$$

which is a multi-variate Gaussian distribution over $\mathbf{f}$. Any Gibbs prior which has a quadratic form may be interpreted as a frequency domain prior using equation (6.21).

6.4 Generic image priors

This section describes some commonly used, simple, generic image priors. Some of these priors will be used to generate the results in the following sections.

The $\|x^2\|$ prior Referring to equation (6.8), and setting $\mathtt{Q}$ equal to some multiple of the identity is equivalent to assuming zero-mean, independent Gaussian distributed pixel values. We shall modify this distribution slightly to use the average image as the mean instead. This allows us to take advantage of the good super-resolution estimate which is provided by the average image, by defining a prior which encourages the super-resolution estimate to lie close to it. The associated Gibbs prior is

$$\Pr(\mathbf{f}) = \frac{1}{Z} \exp\left(-\frac{|\mathbf{f} - \mathbf{f}_{\text{avg}}|^2}{2\sigma_f^2} \right) \tag{6.22}$$

This prior is not actually an MRF, since it involves no spatial correlation between pixels.

Gaussian MRFs When the matrix Q in equation (6.8) is *non-diagonal*, we have a multi-variate Gaussian distribution over **f**, in which spatial correlations between adjacent pixels are captured by the off-diagonal elements. The corresponding MRFs are termed Gaussian MRFs or GMRFs.

As noted in Section 6.2.3, it is common for Q to be defined by the action of some first or second derivative operator L on **f**, such that $\mathtt{Q} \propto \mathtt{L}^\top\mathtt{L}$. For the purpose of our examples, we define a GMRF in which L is formed by taking first-order finite difference approximations to the image gradient over horizontal, vertical and diagonal pair-cliques. For every location $f_{x,y}$ in the super-resolution image, L computes the following finite-differences in the four adjacent, unique pair-cliques :

$$\begin{aligned} d_x &= f_{x+1,y} - f_{x,y} & d_y &= f_{x,y+1} - f_{x,y} \\ d_{xy} &= \frac{1}{\sqrt{2}}(f_{x+1,y+1} - f_{x,y}) & d_{yx} &= \frac{1}{\sqrt{2}}(f_{x+1,y-1} - f_{x,y}) \end{aligned} \tag{6.23}$$

The corresponding Gibbs potentials are $V(C_x) = \gamma d_x^2$, $V(C_y) = \gamma d_y^2$ and so on, where γ is a constant which controls the "peakiness" of the resulting Gibbs distribution. This prior encourages a smooth solution.

In the frequency domain, convolution with a first derivative operator is equivalent to multiplication by $j\omega$. From equation (6.21), we have that

$$\begin{aligned} \mathtt{L}^\top\mathtt{L} &\leftrightarrow P_F(\omega)^{-1} \\ \Rightarrow \omega^2 &\propto P_F(\omega)^{-1} \\ \Rightarrow P_F(\omega) &\propto \frac{1}{\omega^2} \end{aligned}$$

Hence, roughly speaking, this GMRF is assuming that the signal power in **f** decays as $\frac{1}{\omega^2}$. This is in agreement with investigations regarding the statistics of natural scenes [104, 139], where it is observed that the power in the frequency spectrum decays as approximately $\frac{1}{\omega^2}$. However, it has also been noted that it is phase coherence in a signal which is important in defining structure, and in particular discontinuities. The GMRF has no ability to capture this notion.

Schultz and Stevenson [131] suggest a prior based on second derivatives, in which the spatial activity measures are defined over triplet-cliques:

$$\begin{aligned} d_x^2 &= f_{x-1,y} - 2f_{x,y} + f_{x+1,y} & d_y^2 &= f_{x,y-1} - 2f_{x,y} + f_{x,y+1} \\ d_{xy}^2 &= \frac{1}{2}f_{x-1,y-1} - f_{x,y} + \frac{1}{2}f_{x+1,y+1} & d_{yx}^2 &= \frac{1}{2}f_{x-1,y+1} - f_{x,y} + \frac{1}{2}f_{x+1,y-1} \end{aligned}$$

The associated GMRF assumes a $\frac{1}{\omega^4}$ decay in signal power. We will not make use of this particular prior in our examples.

Huber MRFs A common criticism levelled at the GMRF priors is that the associated MAP super-resolution estimates tend to be overly smooth, and that sharp edges, which are what we are most interested in recovering, are not preserved. This problem can be ameliorated by modelling the image gradients with a distribution which is heavier in the tails than a Gaussian. Such a distribution accepts the fact that there is a small, but nonetheless tangible probability of intensity discontinuities occurring.

In a Huber MRF (HMRF), the Gibbs potentials are determined by the Huber function,

$$\begin{aligned} \rho(x) &= \quad x^2 \quad && \text{if } |x| \le \alpha \\ &= 2\alpha\,|x| - \alpha^2 && \text{otherwise} \end{aligned} \tag{6.24}$$

In our examples, the clique statistic x is given by the first derivative difference equations (6.23), and the associated Gibbs potentials are $V(C_x) = \gamma\rho(d_x)$, etc. This prior encourages local smoothness, whilst being more lenient toward step edges than $\rho(\Delta) = \Delta^2$. The associated Gibbs prior is Gaussian close to the origin, becoming Laplacian (double-tailed exponential) in the tails :

$$\begin{aligned} \Pr(x) &= \quad \tfrac{1}{Z}\exp(-\gamma x^2) && \text{if } |x| \le \alpha \\ &= \tfrac{1}{Z}\exp(-\gamma(2\alpha|x| + \alpha^2)) && \text{otherwise} \end{aligned} \tag{6.25}$$

Figure 6.1 shows the Huber potentials function, and the corresponding Gibbs pdf plotted for several values of α. Note that the transition from the quadratic to the linear region maintains gradient continuity. HMRF are an example of convex, but non-quadratic priors.

Schultz and Stevenson use their second derivative scheme with the Huber potentials as a piecewise-smooth image model.

Generalized Gaussian MRFs Bouman and Sauer [21, 124] and Borman *et al.*[19, 20] propose a prior in which the image gradients are modelled by a generalized Gaussian distribution, which is of the form

$$\Pr(x) = \frac{1}{Z}\exp\left(-\frac{x^p}{p\sigma^p}\right) \tag{6.26}$$

When $1 < p < 2$, this produces a heavy-tailed distribution, which the authors claim is also suitable for modelling of piecewise-smooth images. The GGMRF is also a convex, but non-quadratic prior.

Total variation The total variation (TV) norm as a gradient penalty function has become extremely popular in the single-image denoising/deblurring literature in recent years [32, 144, 168, 169]. The Gibbs potentials for this prior are of the form

$$\rho(x) = |x| \tag{6.27}$$

where x is the image gradient statistic within each clique. The corresponding Gibbs pdfs are a double-tailed exponentials. Although convex, there are practical problems when performing optimization involving the TV norm. The gradient of the TV term is

$$\frac{\mathrm{d}\rho}{\mathrm{d}x} = -\int \frac{\nabla.\nabla x}{|\nabla x|} \tag{6.28}$$

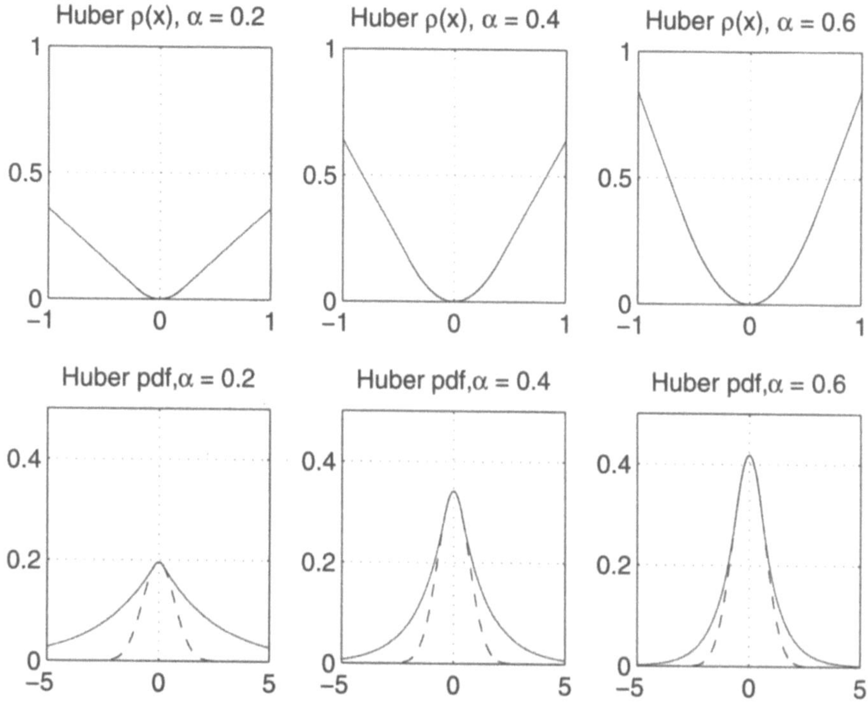

Figure 6.1: (Top) The Huber potential functions $\rho(x)$, plotted for three different values of α. (Bottom) The corresponding Gibbs distributions are a combination of a Gaussian (dashed-line) and a Laplacian distribution.

and hence there is a singularity at $\nabla x = 0$. This is can be problematic for gradient descent or Newton optimization strategies, so the term $|\nabla x|$ is often replaced by $\sqrt{dx^2 + dy^2 + \beta}$, where β is a small constant. An alternative scheme with better global convergence properties is proposed by Chan *et al.* [33].

Weak-membrane MRFs Blake and Zisserman [17] propose a 2D analogue of the weak-string constraint, in which the gradient penalty is

$$\rho(\Delta) = \min(\Delta^2, \alpha^2) \tag{6.29}$$

where Δ is the image gradient. The parameter α controls the step-sensitivity, a threshold beyond which a constant penalty α^2 is applied. $\rho(\Delta)$ can be used as a Gibbs potential given the following modification :

$$\begin{aligned} \rho(\Delta) = &\quad \min(\Delta^2, \alpha^2) \quad \text{if } |\Delta| < \Delta_{\max} \\ = &\quad 0 \qquad\qquad\quad \text{otherwise} \end{aligned} \tag{6.30}$$

where $\Delta_{\max}$ is some maximum plausible gradient, e.g. 255 for an 8-bit image. The corresponding Gibbs distribution is then Gaussian near the origin, and uniform when

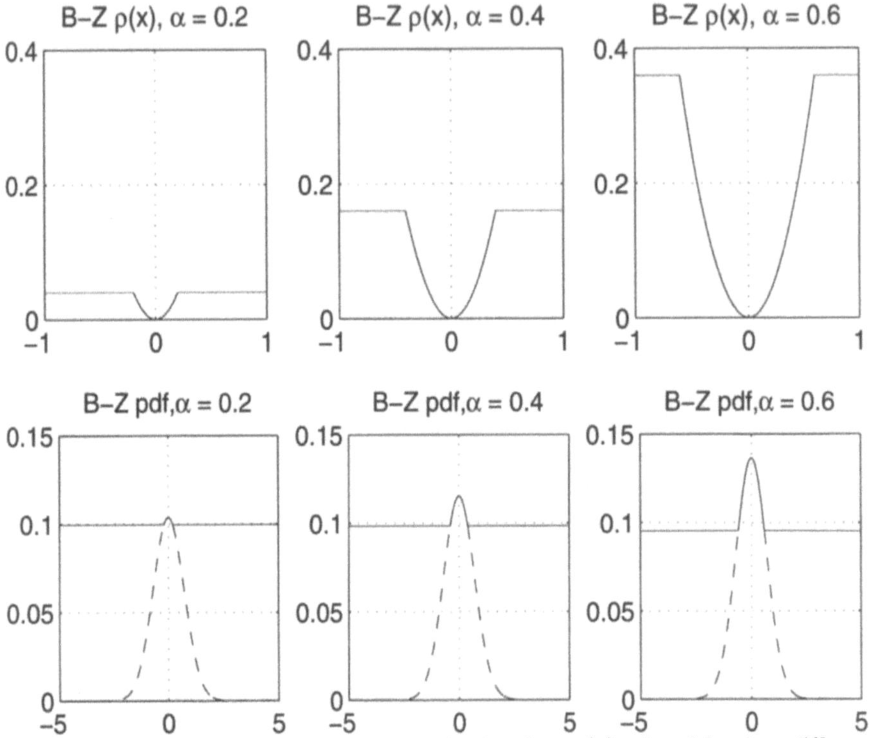

Figure 6.2: (Top) The Blake–Zisserman potential functions $\rho(x)$, plotted for three different values of α. (Bottom) The corresponding Gibbs distributions are a combination of Gaussian and uniform distributions.

$|\Delta| > \alpha$. The potential function and Gibbs distribution is plotted in Figure 6.2 for several values of α.

There are some severe drawbacks with the use of this model. It is both non-convex, has gradient discontinuities, and both zero gradient and zero curvature at high values of Δ. All of these factors make efficient optimization rather difficult. Furthermore it is shown in [21] that images reconstructed using this prior are discontinuous in the data to which the model is being fitted. In other words, small perturbations to the observed data can cause the MAP estimate to "switch modes".

6.5 Practical optimization

The results in the remainder of this chapter use only the $||x||^2$ prior and the first derivative GMRF and HMRF priors. Recalling equation (6.7), the general solution to the MAP estimation problem involving a Gibbs prior,

$$\mathbf{f}_{\text{map}} = \arg\max_{\mathbf{f}} \quad - \sum_{\forall C \in \mathcal{C}} V_C(C) - \frac{1}{2\sigma_n^2} ||\mathbf{Mf} - \mathbf{g}||^2 \tag{6.31}$$

we can write down the MAP estimators for each of the three priors :

$||x||^2$ prior

$$\mathbf{f}_{\text{map}} = \arg\max_{\mathbf{f}} \quad -\frac{1}{2\sigma_f^2}||\mathbf{f} - \mathbf{f}_{\text{avg}}||^2 - \frac{1}{2\sigma_n^2}||\mathtt{M}\mathbf{f} - \mathbf{g}||^2$$

GMRF

$$\mathbf{f}_{\text{map}} = \arg\max_{\mathbf{f}} \quad -\gamma||\mathtt{L}\mathbf{f}||^2 - \frac{1}{2\sigma_n^2}||\mathtt{M}\mathbf{f} - \mathbf{g}||^2$$

HMRF

$$\mathbf{f}_{\text{map}} = \arg\max_{\mathbf{f}} \quad -\gamma\rho(\mathtt{L}\mathbf{f}) - \frac{1}{2\sigma_n^2}||\mathtt{M}\mathbf{f} - \mathbf{g}||^2$$

Each prior has associated with it some variance-like parameter which determines the spread of the distribution: σ_f^2 in the case of the $||x||^2$ prior, γ in the case of the first-derivative GMRF and HMRF. Typically, these parameters are not known *a priori*. Neither is the image noise σ_n, for which it is generally rather hard to obtain an accurate estimate. Consequently, we adopt the common practice of absorbing both of these parameters into a single parameter λ which is the ratio of the two. The MAP estimate using the $||x||^2$ prior becomes

$$\mathbf{f}_{\text{map}} = \arg\max_{\mathbf{f}} \quad -\lambda||\mathbf{f} - \mathbf{f}_{\text{avg}}||^2 - ||\mathtt{M}\mathbf{f} - \mathbf{g}||^2$$

The others follow similarly. We return to the problem of selecting an optimal value for λ in Section 6.7.

For the $||x||^2$ and GMRF priors, the cost function is quadratic, and hence optimization may be efficiently performed by conjugate gradient ascent. The HMRF prior, on the other hand, introduces a non-quadratic term into the cost function, and hence the solution requires a general, unconstrained non-linear solver. For this purpose, we use a damped Newton method which is directly analogous to the Levenberg–Marquardt algorithm for non-linear least-squares problems which was used in the N-view homography estimator of Chapter 4. The difference is that we must use the full Hessian of the non-linear cost function, rather than the Gauss–Newton approximation which may be used in non-linear least-squares problems. The method is described in more detail in Appendix A.

6.6 Sensitivity of the MAP estimators to noise sources

In this section, we use our synthetic image sequences to examine the noise robustness properties of the $||x||^2$, GMRF and HMRF priors described in the previous section.

6.6.1 Exercising the prior models

To get some idea of the effect that each prior has on the reconstructed image, we first perform a reconstruction from 20 noise-free images, varying λ from very small to very large. Figures 6.3 and 6.4 show reconstructions using the simple $||x||^2$ and

GMRF priors respectively. Both priors produce a similar result. They are not good at preserving step edges as λ increases. It is interesting to note that the $||x||^2$ prior, although it does not directly impose spatial correlation between pixels in the solution, still tends to produce a smoothed solution for large values of λ.

Figures 6.5, 6.6 and 6.7 show results using the HMRF prior for three different values of α. As α decreases, the step edge preserving property of the HMRF prior becomes apparent. Note though that this prior also exhibits a kind of hysteresis effect; having allowed a step to occur, it is reluctant to allow another nearby. Hence the highest frequency bars are turned into solid grey blocks, albeit with sharp edges.

6.6.2 Robustness to image noise

Figures 6.8, 6.9 and 6.10 show reconstructions using the three MAP estimators applied to 20 noisy synthetic images. The reconstructions are all performed at $2\times$ zoom, and the synthetic images are all degraded by quite severe additive Gaussian, $\sigma = 10$ grey-levels. The behaviour of the $||x||^2$ and GMRF priors is much as expected; when λ is small, the solution is much like the MLE, and consequently noisy. When λ is very large, the solution is not noisy, but is overly smooth. Somewhere in between lies the optimal value of λ which balances noise against smoothness in the solution.

The reducing effect that α has on the penalty term in the HMRF prior generally means that λ must be higher in order to achieve a similar regularizing effect as the other two MAP estimators. The edge preserving characteristic of the estimator becomes apparent when λ is large and α small, the reconstructions taking on a cartoon-like appearance. This prior is particularly suitable for reconstructing scenes in which the intensities are quantized into a small number of distinct levels, such as images containing text.

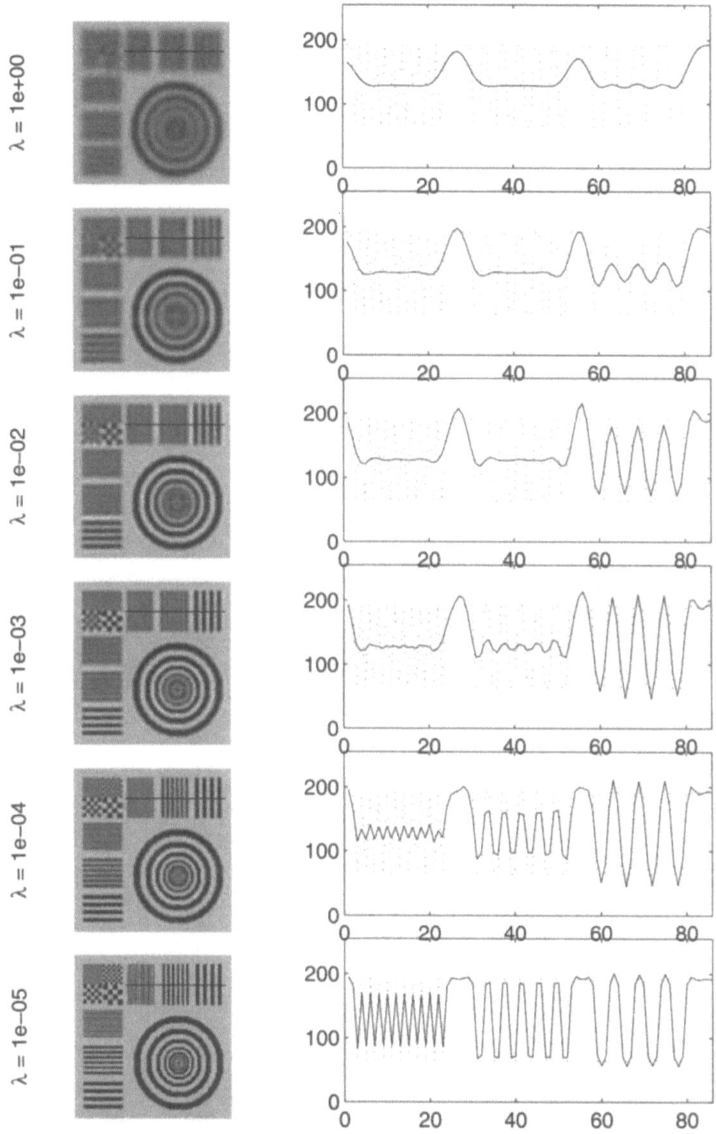

Figure 6.3: Reconstructions at $2\times$ zoom from 20 noise-free synthetic images using the simple $\|x\|^2$ prior, as λ varies. This prior cannot preserve step edges, which become smoothed.

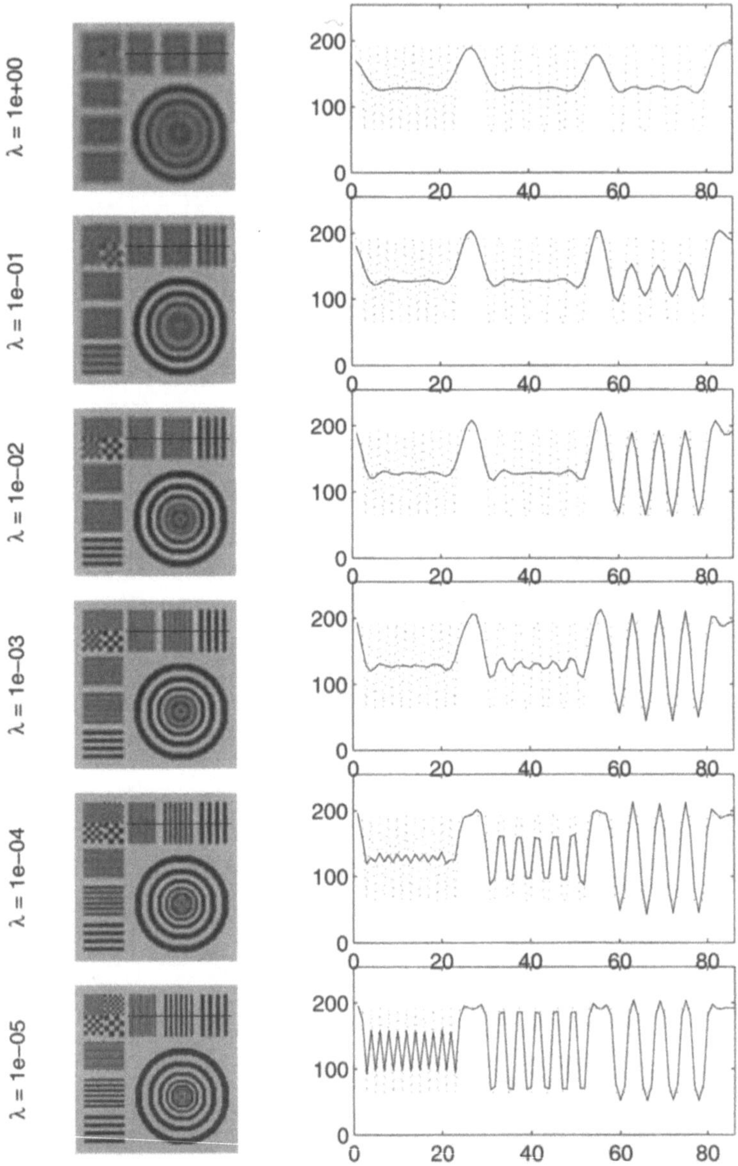

Figure 6.4: Reconstructions at 2× zoom from 20 noise-free synthetic images using the simple GMRF prior, as λ varies. This prior is also unable to preserve step edges.

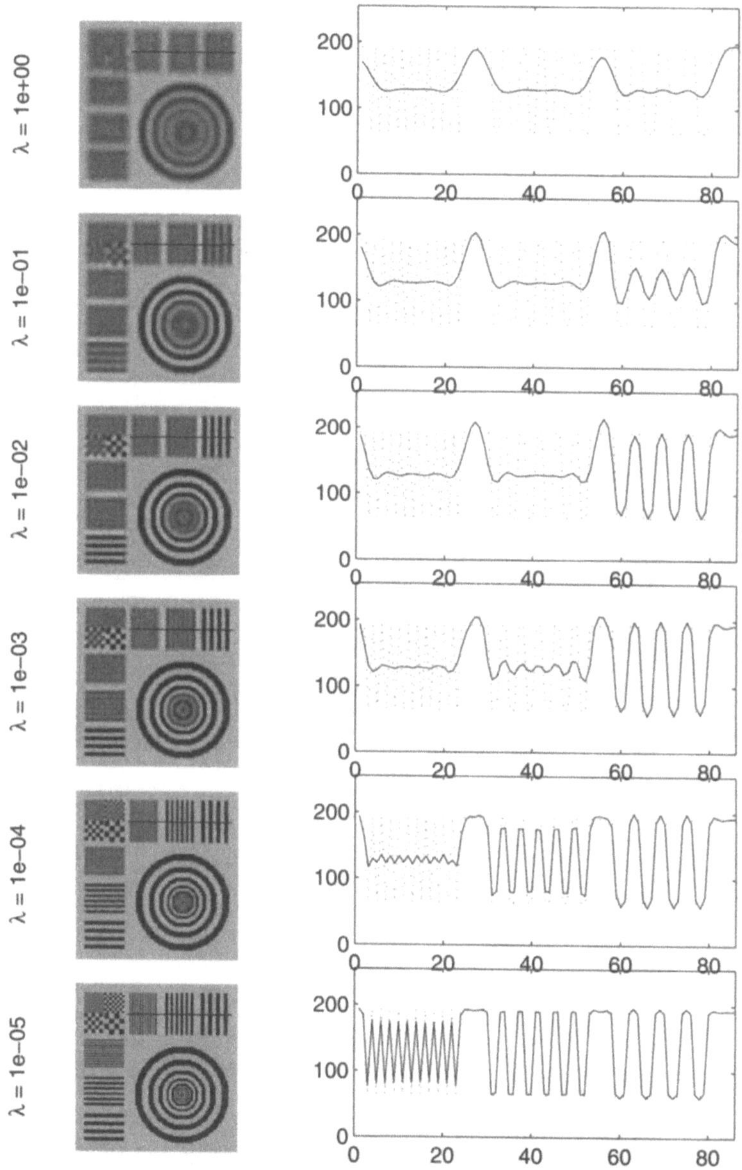

Figure 6.5: Reconstructions at $2\times$ zoom from 20 noise-free synthetic images using the simple HMRF prior, as λ varies with fixed $\alpha = 0.1$. As α decreases, we expect this prior to be much better at preserving step edges than the $||x||^2$ and GMRF priors.

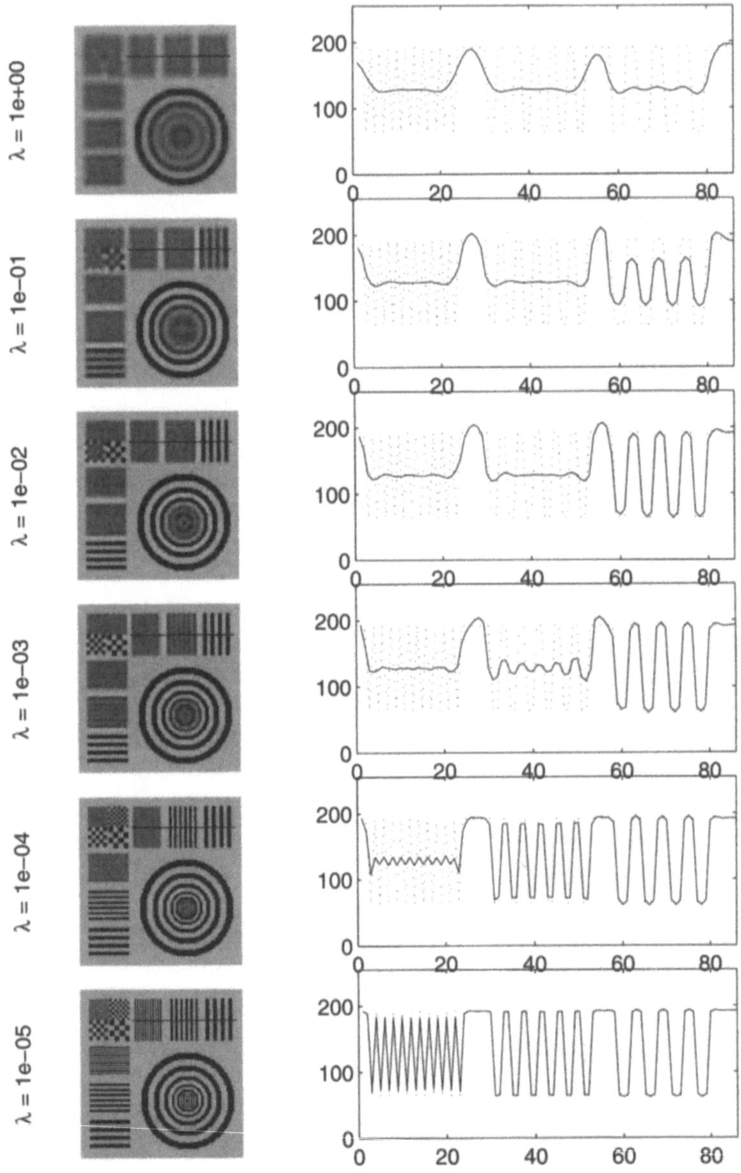

Figure 6.6: As Figure 6.5, but with $\alpha = 0.05$.

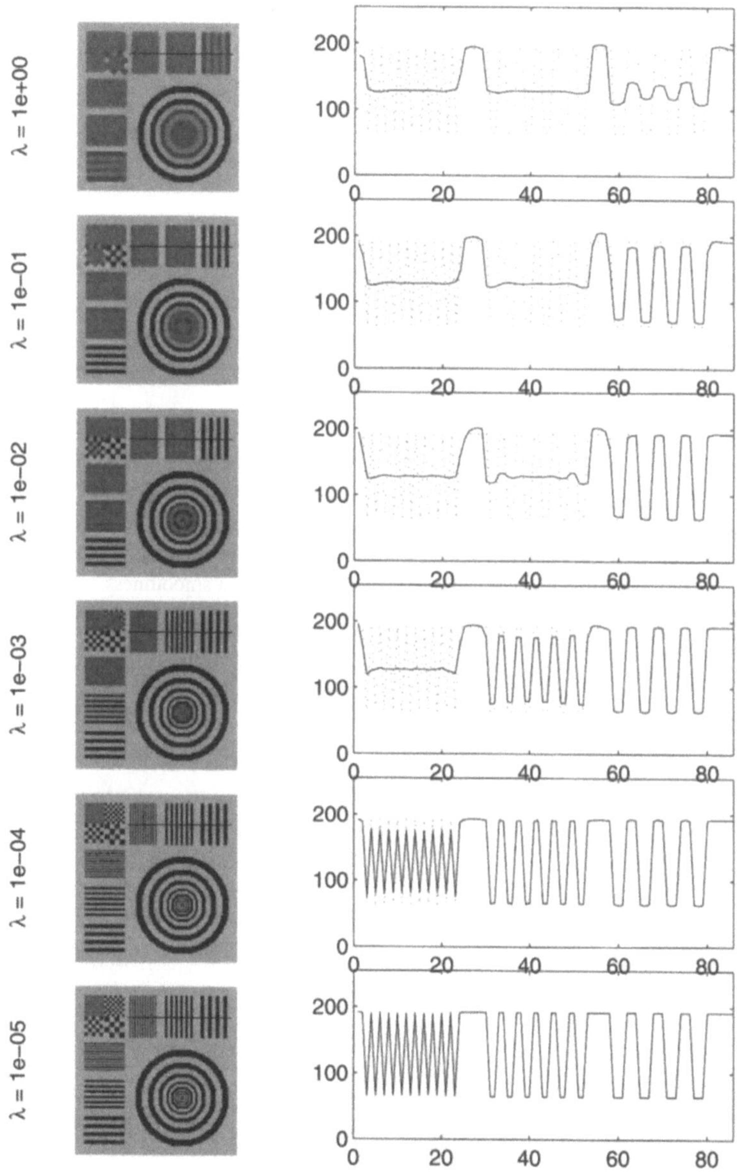

Figure 6.7: As Figure 6.5, but with $\alpha = 0.01$.

Figure 6.8: Reconstructions using the $\|x\|^2$ MAP estimator at $2\times$ pixel-zoom using the synthetic images described in the text. A small λ produces a solution close to the MLE, which is corrupted by noise. A very large λ reduces noise effectively, but produces an overly smooth solution. The optimal value balances reconstruction noise against smoothness.

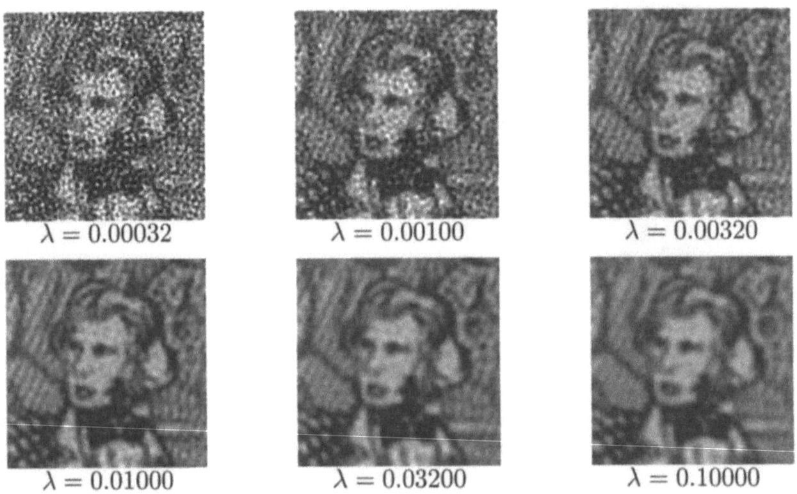

Figure 6.9: Reconstructions using the GMRF-MAP estimator. The behaviour follows the same pattern as the $\|x\|^2$ prior in Figure 6.8.

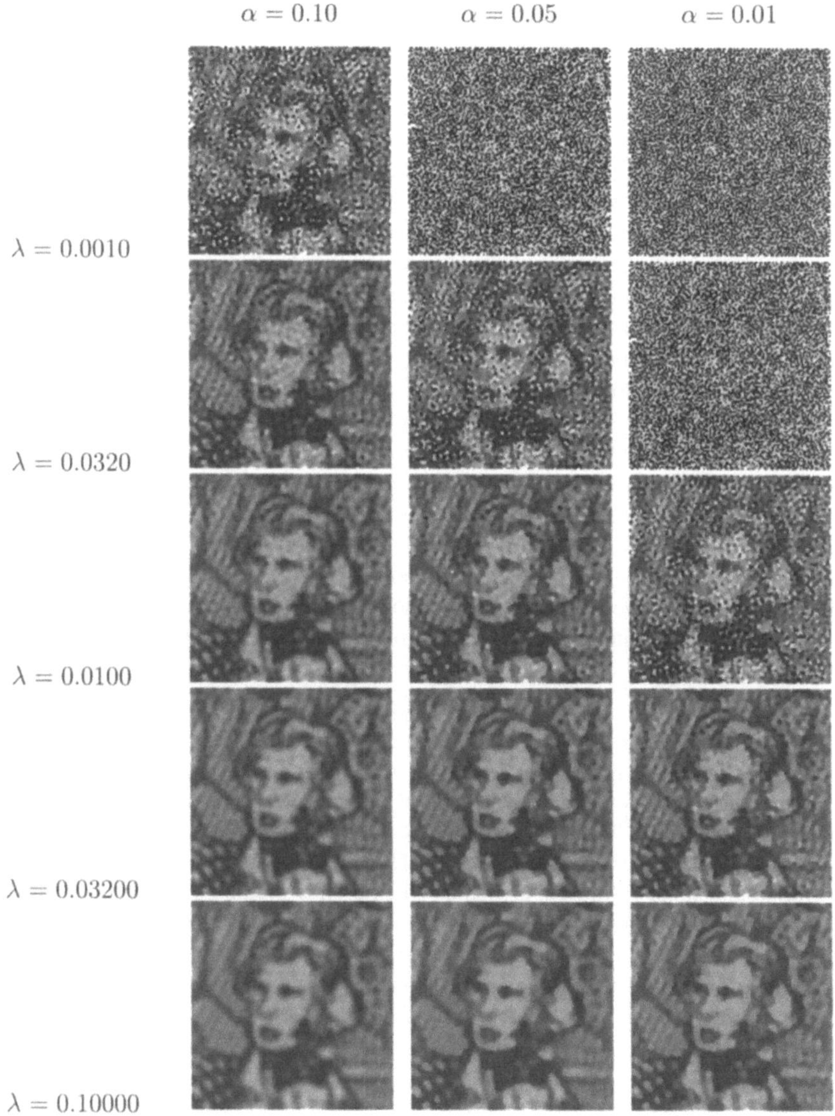

Figure 6.10: Reconstructions using the HMRF-MAP estimator for varying λ and α. The value of λ must generally be higher than that used in the $||x||^2$ and GMRF estimators in order to achieve a similar regularizing effect. In the bottom-rightmost images, the edge preserving characteristic of the HMRF becomes clear, as the images take on a rather cartoon-like appearance.

6.7 Hyper-parameter estimation by cross-validation

In this section, we demonstrate how a cross-validation technique may be used to estimate the optimal value of the prior influence parameter λ.

The cross-validation scheme employed here is an example of "hold-out" cross-validation. The basic principle is to divide the observed data into two sets, compute a MAP estimate using only one of the sets, and then use this estimate to compute the cross-validation residual error over the second set. The process is repeated, adjusting the hyper-parameter, so as to minimize the cross-validation residual. We shall refer to the two sets as the *fitting* set and the *validation* set.

In this context, the data set is divided by holding-back every n^{th} image, and computing the super-resolution MAP estimate, given some value of λ, using the remaining images. The estimate is then projected into each image in the validation set, and the total RMS reprojection error computed.

The intuition as to why this works is as follows. When λ is very large, the super-resolution estimate is overly smooth. Consequently, the reprojection error with respect to any image in the sequence is large. Initially, as λ gets smaller, both the fitting residual and the validation residual decrease, and the MAP estimate improves. However, when λ is very small, the MAP estimate tends toward the ML estimate and is *over-fitted* to the data, appearing extremely noisy. In this case, the fitting residual is artificially small, smaller than the actual noise level in the images, but the validation residual is again large. Hence, the validation residual has a minimum with respect to λ. In a fully automated system, this minimum could be sought by an iterative root-finding algorithm.

The choice of validation set It is important that the set of images held back for validation purposes is a good "cross-section" of the entire set of images. A set of mutally very similar images will not provide a good validation set. The method employed here of choosing every n^{th} image is an attempt to obtain such a cross-section. Of course, the larger the size of the validation set, the fewer images are left with which to actually compute the super-resolution estimate. One possible method by which both of these issues might be addressed is to use several small, randomly chosen validation sets. Each validation set is then held back in turn, with all of the remaining images going into the fitting set. The optimization task is then to choose the hyper-parameter so as to minimize the total cross-validation residual across *all* of the validation sets.

A synthetic experiment For this experiment, 25 images were generated from the "Text" ground-truth image undergoing projective motion. The point-spread function σ_{psf} was 0.7, the down-sampling ratio was 3, and zero-mean, additive Gaussian noise with $\sigma_n = 5$ grey-levels was added to the images. Every fifth image was held-back. The MAP estimator used the $\|x\|^2$ prior.

Figure 6.11 shows MAP estimates computed using the 20 remaining images, at $3\times$ pixel-zoom, as $\log_{10}\lambda$ varies from -1 to -5. As expected, for a large value of λ, the estimate is overly smooth. When λ is very small, the estimate is very noisy. Subjectively, we might say that $\log_{10}\lambda = -3$ looks about the best.

Figure 6.12(a) shows the variation of the fitting residual with respect to λ. As expected, the error decreases monotonically with decreasing λ, and offers no help in choosing an optimal value. Figure 6.12(b) shows the corresponding variation of RMS reconstruction error compared with the ground-truth. The minimum appears at around $\log_{10} \lambda = -3$. Figure 6.12(c) shows the variation of validation residual with respect to λ. A close-up of graphs (b) and (c) is shown in Figure 6.13. The minimum of the validation residual occurs in almost exactly the same place as the minimum ground-truth error, and hence provides an excellent guide as to the optimal value of λ. Figure 6.14 shows one of the low-resolution input images, the average image, and the MAP estimate computed using $\log_{10} \lambda = -3$.

A real example To demonstrate the cross-validation approach applied to a real example, we use the "Wedge" sequence of 25 JPEG images, first shown in Section 5.13. Again, every fifth image is held-back. The MAP estimate is computed using the first-derivative GMRF prior, at $3\times$ pixel-zoom. The point-spread function σ_{psf} for this sequence is 0.4.

Figure 6.15 shows MAP estimates as $\log_{10} \lambda$ varies from 0 to -4. For clarity, only a 100×100 pixel region from the full 300×300 pixel reconstruction is shown. Figure 6.16(a) demonstrates the monotonic decrease of the fitting residual with respect to λ. Figure 6.16(b) shows the validation residual, which has a minimum at around $\log_{10} \lambda = -2$. Figure 6.17 shows one of the low-resolution input images, the average image, and the MAP estimate using this value of λ. The result is clear, without being overly noisy, exactly as desired.

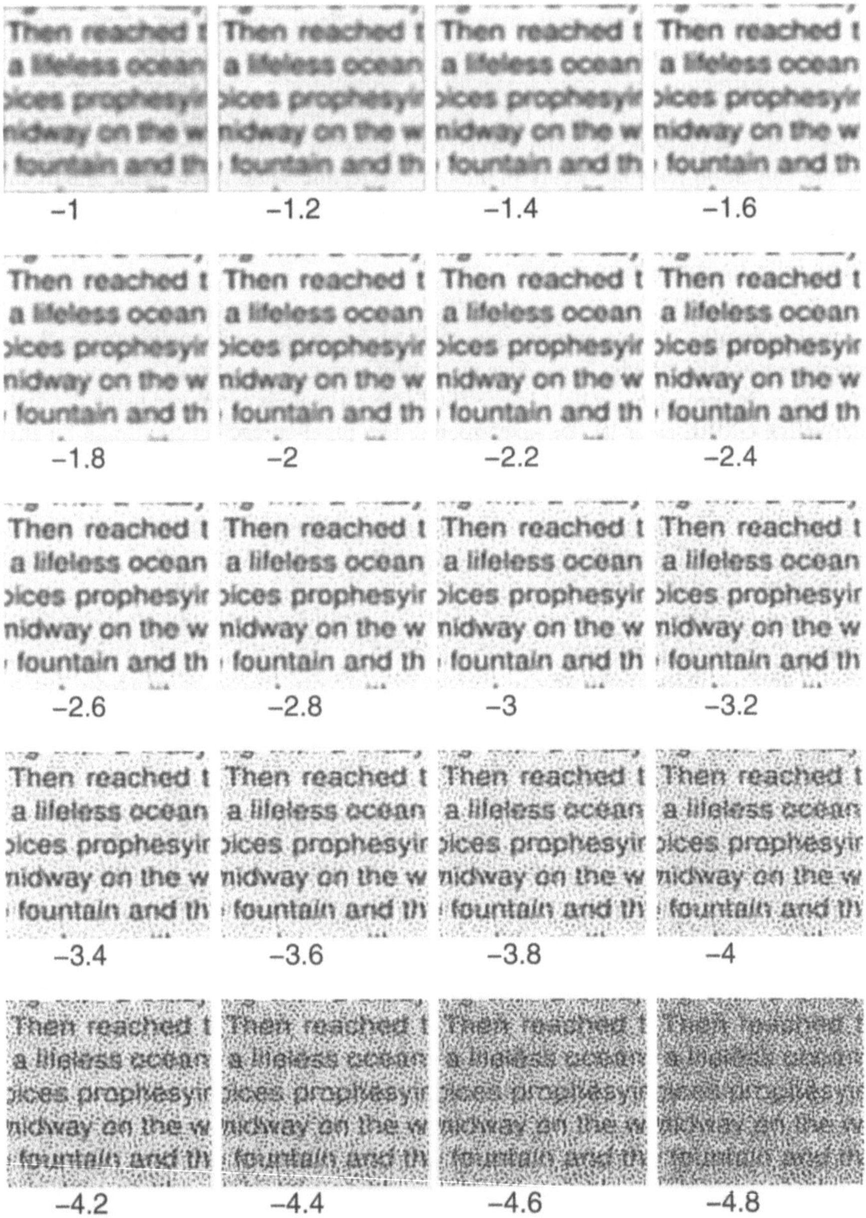

Figure 6.11: MAP estimates computed using 20 noisy, synthetic images (the synthesis parameters are described in the text). The $\|x\|^2$ prior was used. $\log_{10} \lambda$ varies from -1 to -5. The subjective optimal value appears to be around $\log_{10} \lambda = -3$.

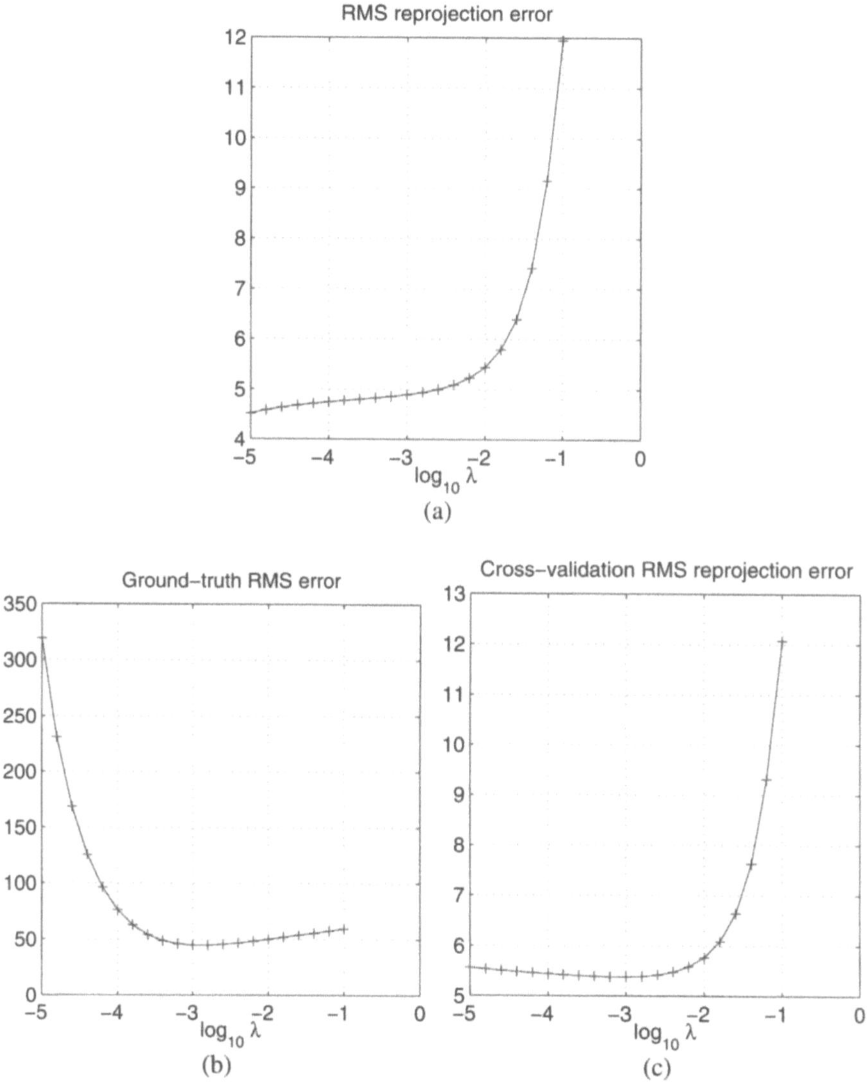

Figure 6.12: (a) The fitting residual decreases monotonically with respect to λ. (b) The ground-truth reconstruction error has a mininum at approximately $\log_{10} \lambda = -3$. (c) The minimum of the validation residual is also at $\log_{10} \lambda = -3$. A close-up view of graphs (b) and (c) is shown in Figure 6.13. The validation residual therefore provides an excellent estimate of the optimal value of λ.

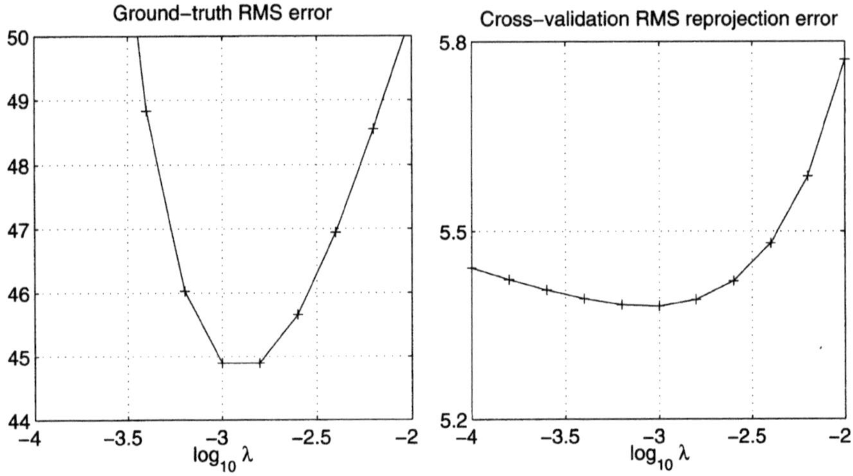

Figure 6.13: Close-up views of graphs (b) and (c) from Figure 6.12.

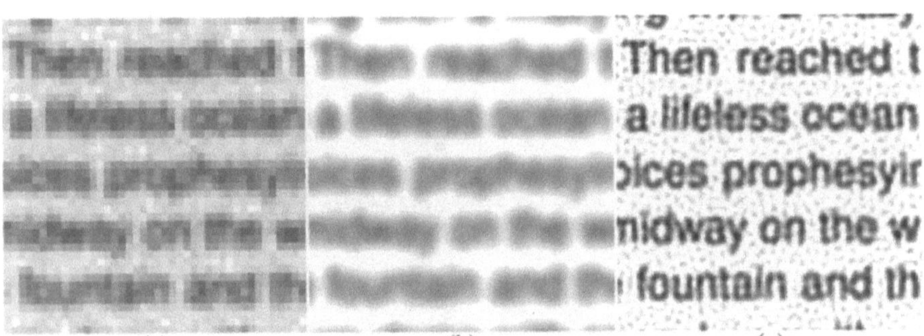

Figure 6.14: (a) One of the 20 noisy, low-resolution input images. (b) The average image. (c) The super-resolution $\|x\|^2$ prior MAP estimate at $3\times$ pixel-zoom, computed using $\log_{10}\lambda = -3$.

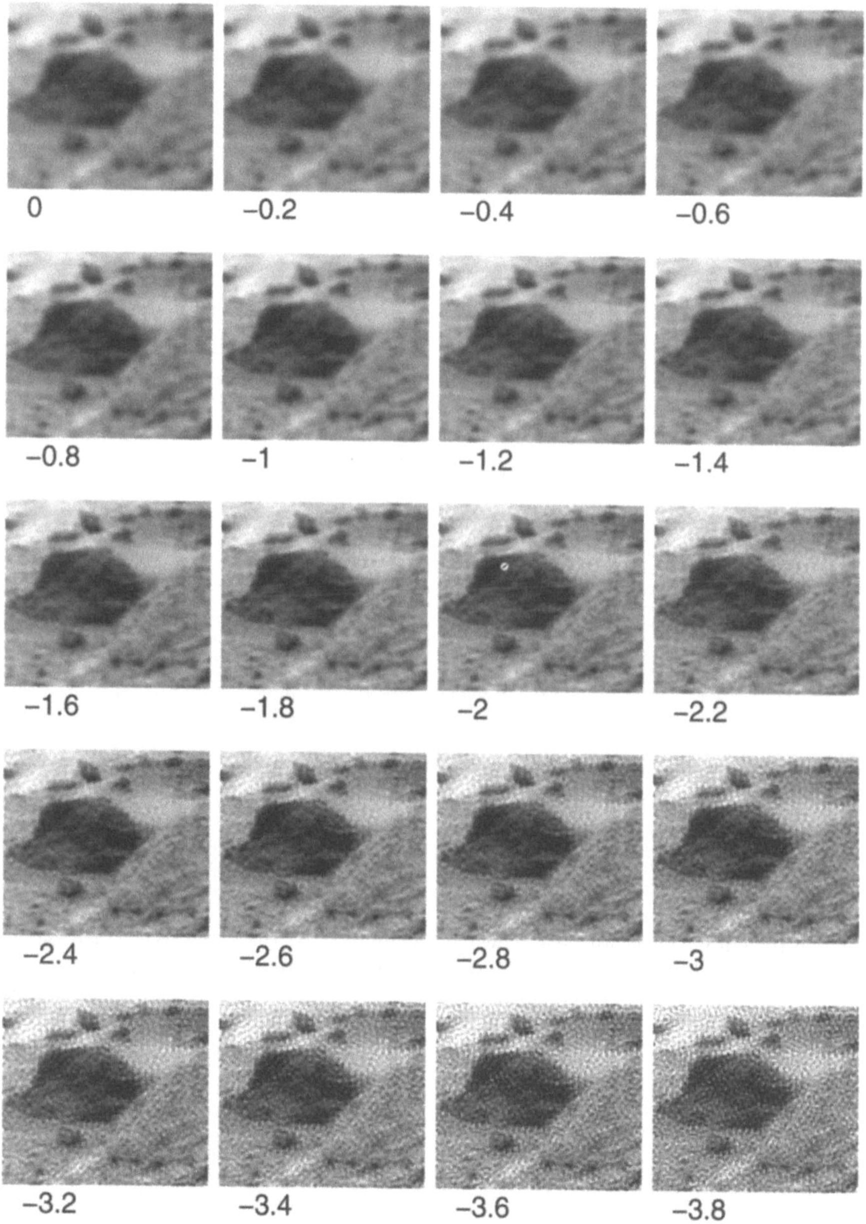

Figure 6.15: Reconstructions at $3\times$ pixel-zoom using the GMRF-MAP estimator prior applied to the "Wedge" sequence of 25 images. $\log_{10}\lambda$ varies as indicated. Only 100×100 pixel regions from the full 300×300 pixel reconstructions are shown. The subjective optimum occurs around $\log_{10}\lambda \sim -2$.

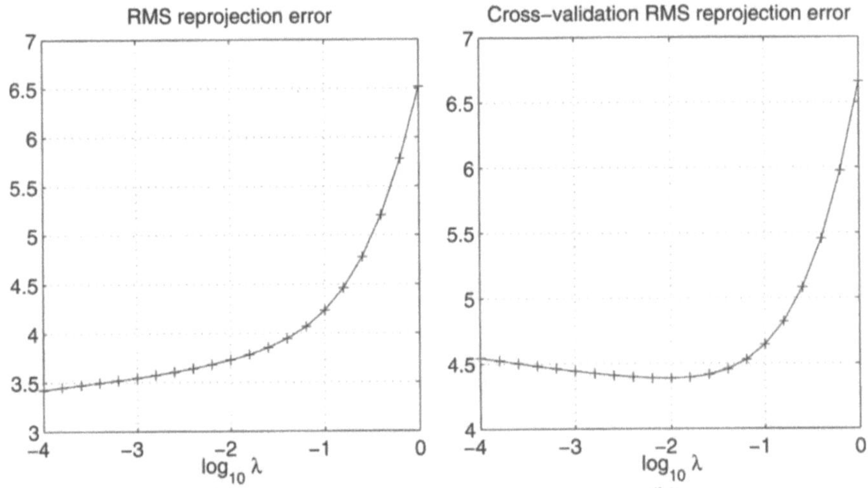

(a) (b)

Figure 6.16: For the reconstructions shown in Figure 6.15: (a) The fitting residual monotonically decreases with respect to λ. (b) The validation residual has a minimum at $\log_{10} \lambda = -2$, which agrees with the subjective optimum.

Figure 6.17: (a) One of the 25 low-resolution images from the "Wedge" sequence (shown at $3\times$ zoom using bicubic interpolation). (b) The average image. (c) GMRF-MAP reconstruction at $3\times$ pixel-zoom, with the value of λ selected using the cross-validation method ($\lambda = 0.01$). The result is clear and not noisy.

6.8 Gallery of results

Figures 6.18 to 6.20 show results generated using the methods described in this chapter applied to the same sequences as were shown in Section 5.13. The best ML result is also shown in each case for the purpose of comparison. All of the MAP reconstructions are obtained at 3× pixel-zoom ratio. The value of σ_{psf} for each sequence is the same as was quoted in Section 5.13.

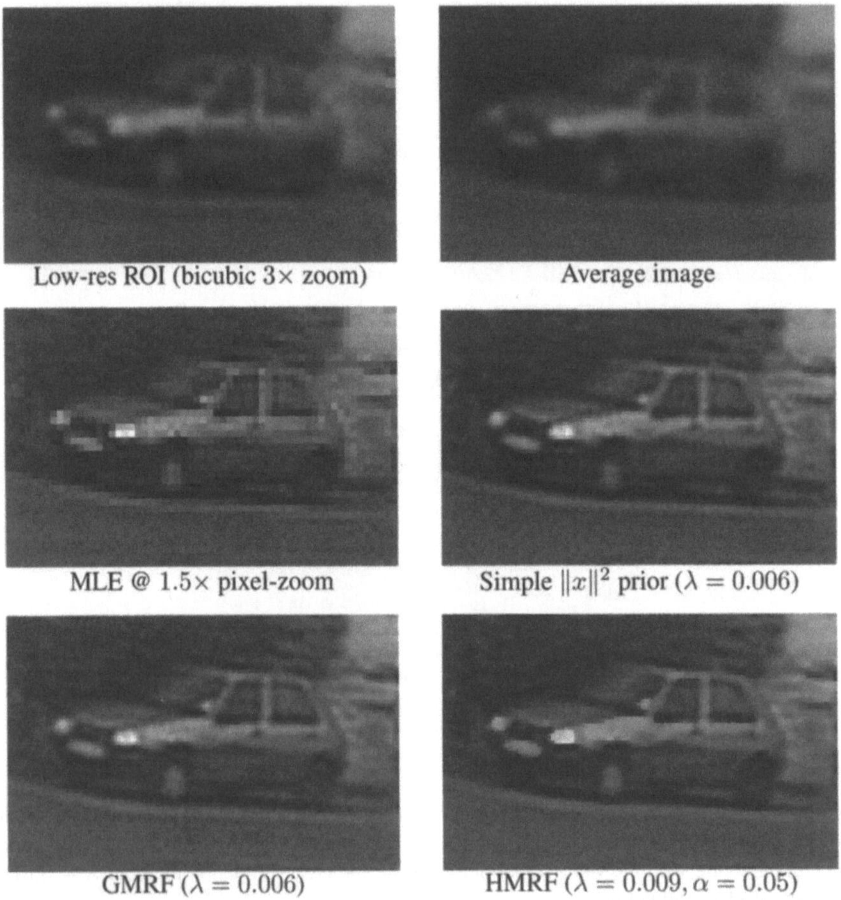

Figure 6.18: The three MAP results, reconstructed at 3× pixel-zoom, all show more convincing detail than the ML reconstructed, especially around the door handles and wing mirror. The $\|x\|^2$ and GMRF priors produce similar results, but note the sharp edges around the windows and headlights in the HMRF reconstruction. The level of detail in the reconstructions compared with the low-resolution images is very apparent.

Low-res ROI (bicubic $3\times$ zoom)

Average image

MLE @ $1.5\times$ pixel-zoom

Simple $||x||^2$ prior ($\lambda = 0.004$)

GMRF ($\lambda = 0.003$)

HMRF ($\lambda = 0.01, \alpha = 0.05$)

Figure 6.19: Again, the superiority of the MAP results over the MLE is quite convincing. The HMRF estimator is particularly effective at reconstructing the sharp edges of the letters in this example. The text is clearly legible, which is certainly not the case in the low-resolution input.

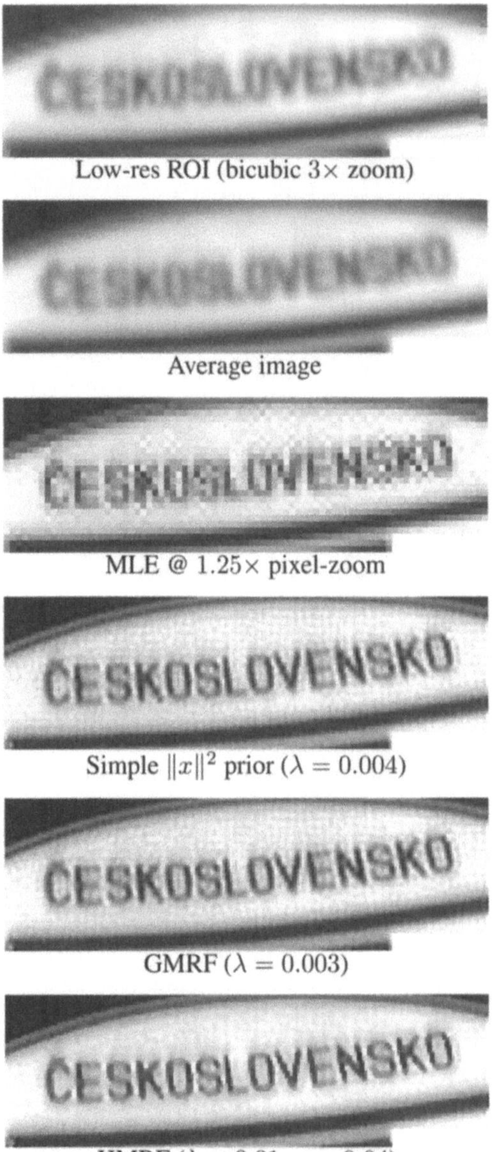

Figure 6.20: For this sequence the ML estimator starts to break down at only $1.25\times$ pixel-zoom. Considering the small number of images in this sequence (only 10), the quality of the MAP reconstructions compared with that of the low-resolution input is fairly remarkable. Again, the edge-preserving property of the HMRF is particularly suited to this example.

6.9 Super-resolution "user's guide"

The following paragaphs summarize the imaging situations to which each of the super-resolution algorithms discussed so far is best suited.

ML estimator This estimator is the simplest to implement, requiring only a large-scale, quadratic optimization algorithm such the preconditioned conjugate gradients (PCG) method in order to perform the estimation. It has the drawbacks of being extremely sensitive to observation noise and registration error, and is only useful for very low zoom ratios (typically $< 2\times$).

$||x||^2$ **prior** Also requires only quadratic optimization. It is superior to the ML estimator in terms of noise and mis-registration tolerance, and has the advantage that it achieves this robustness without explicitly imposing any spatial smoothness on the super-resolution estimate. Suitable for use with zoom ratios $> 2\times$. It is hard to imagine a scenario in which the ML estimator would be preferred over this one.

GMRF prior Requires only quadratic optimization. It also achieves far greater robustness than the ML estimator, but does so by imposing a penalty on spatial "non-smoothness" in the super-resolution estimate. In practice, this algorithm performs very similarly to the $||x||^2$ prior, and either is a good choice for performing super-resolution restoration of general scenes.

HMRF prior This estimator requires a large-scale, non-linear optimizer such as a globalized Newton method. It also achieves excellent noise robustness, achieved by imposing *piecewise* spatial smoothness on the super-resolution estimate. This gives it the ability to preserve and enhance edges. This estimator is best suited to restoration of approximate piecewise constant intensity profiles, such as images of text.

6.10 Summary

In this chapter, we have investigated the use of Bayesian prior image models as a regularizing influence in super-resolution restoration. Several "generic" image priors have been described and their properties exercised by empirical investigation using synthetic images. A novel algorithm has been described which uses cross-validation to set the level of influence that the prior model has on the MAP solution in order to minimize reconstruction error without introducing excessive smoothness. The algorithm has been shown to produce near-optimal results on both real and synthetic examples. Finally, the performance of various of the MAP estimators was demonstrated by application to real image sequences. In each case, the MAP estimators were seen to outperform the ML estimator in terms of reconstruction quality, and ability to restore images at greatly increased pixel densities.

In the next chapter, we investigate whether the performance of super-resolution algorithms can be improved still further by the use of image models which are tuned to a specific class of image.

Chapter 7

Super-resolution Using Sub-space Models

7.1 Introduction

In the previous chapter we demonstrated that generic MRF prior models can go a long way towards improving the performance of our super-resolution estimators. However, the generality of these priors is also their weakness, and in computing a super-resolution MAP estimate, there is a trade-off to be made between reduced noise and excessive smoothness.

In this chapter we introduce the use of compact image models that are tuned to particular classes of image. In certain cases, the range-space of these models may be learnt from training images. We consider in particular the case of super-resolution reconstruction of text and face images. In the first case, we show that a simple constrained model can give better results than the MAP-MRF estimators, whilst avoiding the need to impose spatial correlation. In the second case, we use a simple face model analogous to the "IdentiKit" method often used to compose images of police suspects. The model is composed from compact models of key facial features – the eyes, nose, mouth and cheek areas – which are learnt from training data using principal-components analysis. The power of this compact representation is demonstrated as we derive an ML estimator and two MAP estimators based on the model. In both cases, constraining the super-resolution estimate to lie in or near a low-dimensional, highly problem-specific sub-space greatly improves the conditioning of the problem. The MAP results typically exceed those possible with generic image priors.

The methods presented here are similar in spirit to those proposed by Baker and Kanade [10] who also examine the use of spatially varying priors for face images. The novelty over their approach is two-fold: first, the low resolution images need not be at the same resolution and indeed the resolution can vary across the image – this generalization is essential in the case that the low resolution images are related by a transformation more general than 2D pure translation; second, a generative model is used throughout.

In Section 7.2 we demonstrate the power of a simple constrained model applied to the reconstruction of text images. Section 7.3 describes the PCA-based face model, and Section 7.4 describes an ML estimator and two MAP estimators based on this model. In Section 7.5, the behaviour of the estimators is analysed using synthetic images. Finally, Section 7.6 presents results of the estimators applied to real image sequences.

7.2 Bound constraints

The motivation for this chapter comes from the success of a very simple constrained model applied to the super-resolution reconstruction of text image sequences. Although not properly explored in the super-resolution literature, it is widely recognized in the single-image restoration literature that placing *hard constraints* on the individual pixel intensities, which restrict the solution to some sub-space of the full image space, can give excellent results without the need for a spatial prior. Examples of such constraints are *non-negativity* and upper/lower bound constraints on pixel values.

As a motivating example, Figure 7.1 compares super-resolution estimates computed by both unconstrained and bound constrained ML estimators using synthetic images. The synthetic images are generated from the "Text" ground-truth images, using a $\sigma_{\text{psf}} = 0.7$ and down-sampling ratio $S = 3$. The super-resolution images are therefore constructed at $3\times$ zoom. Three different levels of Gaussian noise is added to the synthetic images. Even for these tiny levels of noise, the unconstrained ML estimate is completely corrupted by reconstruction error. The bound-constrained estimate on the other hand is unaffected and of high-quality.

Figure 7.2 compares the performance of the bound constrained estimator with results obtained using MAP estimators. In this case the noise on the synthetic images is a much more realistic $\sigma = 5$ grey-levels. The constrained result is clearly superior to both MAP results. The lack of any imposed spatial correlation means that the constrained estimator is able to generate sharper results than the MAP estimators using spatial MRF priors.

Figure 7.3 compares ML, HMRF and bound-constrained reconstructions of a real image sequence – the "Czech" sequence seen in Section 5.13. The reconstruction is performed at $1.75\times$ pixel zoom. The ML estimate is badly corrupted, whilst the bound-constrained estimate is comparable in quality to the MAP estimate using the first-derivative HMRF spatial prior.

Optimization The method used to compute the bound-constrained estimates is More and Toraldo's "Gradient-projection conjugate gradient" (GPCG) algorithm [101]. This method is efficient for the solution of large-scale quadratic problems when the number of constraints is very large, as here. The number of iterations required is typically fewer than that required for convergence of the unconstrained ML estimator.

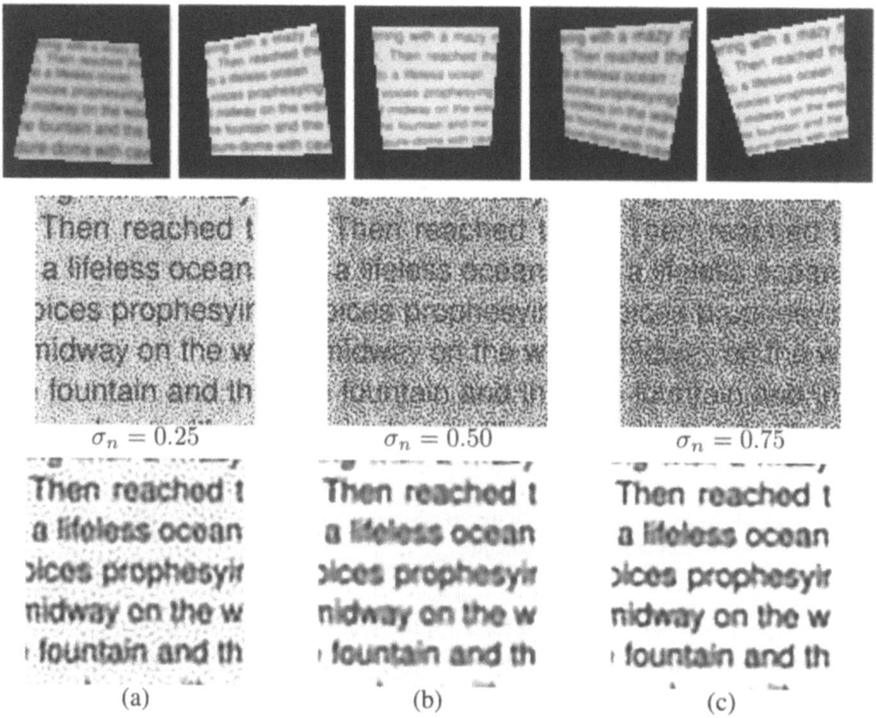

Figure 7.1: (Top) 5 of 30 synthetic perspective images generated from the "Text" ground-truth image. The images have been sub-sampled by a factor of 3 compared with the ground-truth. (Middle) ML reconstructions at $3\times$ zoom for three different levels of Gaussian noise (in grey-levels) added to the low-resolution images. (Bottom) The same ML reconstructions with a 0/255 lower/upper bound on the pixel intensities. The reconstruction error has been eliminated in the constrained cases.

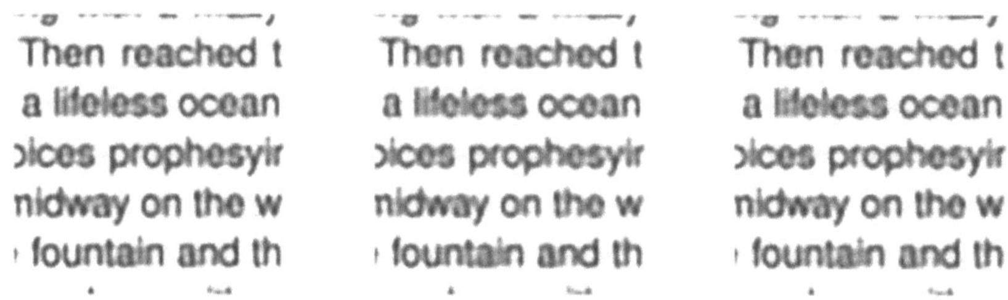

Figure 7.2: Reconstructions computed from the same 30 images used in Figure 7.1, but with a much higher level of Gaussian noise added ($\sigma = 5$ grey-levels). (a) Simple $\|x^2\|$ Tikhonov prior with $\lambda = 0.005$. (b) First-derivative HMRF prior with $\lambda = 0.005, \alpha = 0.1$. (c) Bound-constrained. The constrained estimator out-performs both MAP estimators on this problem.

7.3 Learning a face model using PCA

In this section the image is modelled using a principal components basis computed from training images at the target resolution. Given a set of images, the variation in

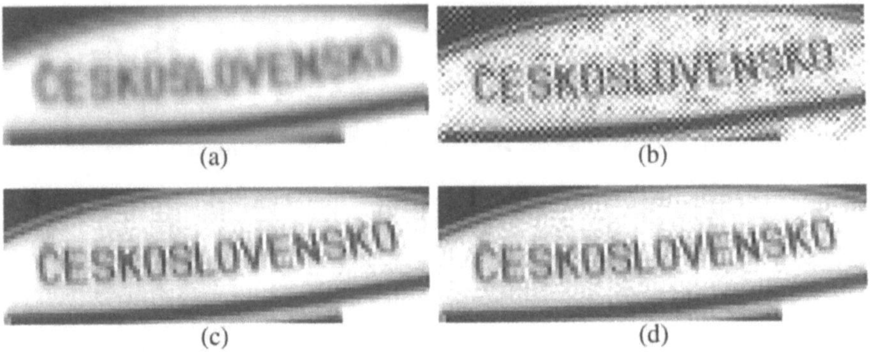

Figure 7.3: Super-resolution reconstructions of the "Czech" sequence of 10 real images, first shown in Section 5.13, at 1.75× pixel-zoom. (a) The region of interest in the reference low-resolution image (shown with 1.75× bicubic zoom). (b) The ML estimate. (c) The HMRF estimate with $\lambda = 5 \times 10^{-3}, \alpha = 5 \times 10^{-2}$ (d) The bound-constrained ML estimate. The ML estimate is badly corrupted, whilst the bound-constrained estimator produces a clean, clear result close to that of the HMRF estimator.

the signals is optimally modelled for a given number of principal components. This representation is applicable to signals of a particular class where the aim is to model the within-class variation.

The use of PCA sub-space priors provides an interesting middle ground between the generic local priors previously used, and the type of priors based on sampling exemplars from training images (e.g. for homogeneous texture generation [51]) which have been applied to super-resolution by Candocia and Principe [26], and Freeman and Pasztor [61].

The image will be modelled by its PCA components as $\hat{\mathbf{f}} = \mathtt{V}\mathbf{y} + \boldsymbol{\mu}$, where $\mathtt{V}$ represents the set of PC basis vectors and $\boldsymbol{\mu}$ is the average of the training images. The aim is a low-dimensional representation where the dimension of the parameter vector $\mathbf{y}$ is far less than the number of pixels in the image (the dimension of $\mathbf{f}$). This representation is applied here to registered face images. It is known from extensive use of PCA since at least [166] that a compact representation can be achieved in this case.

Rather than learn the PC for the entire image, the face is divided into four key facial regions: the eyes, nose, mouth and cheek areas; and a separate PCA basis is learnt for each feature. The intuition is that these regions are relatively uncorrelated, and that by considering small regions, better models can be learnt than would be by performing PCA on the whole face.

The region segmentation is shown in Figure 7.4. To learn the model, PCA analysis is applied to 160 face images which have been geometrically registered using a similarity transformation. The registration is computed from manually selected points on the eyes and mouth. Both male and female faces are used in the training set. To increase the number of training samples, each face is flipped symmetrically around the y-axis, providing a total of 320 faces. Some of the training faces are shown in Figure 7.5. The average eye, nose and mouth features and the first few "eigen-features"

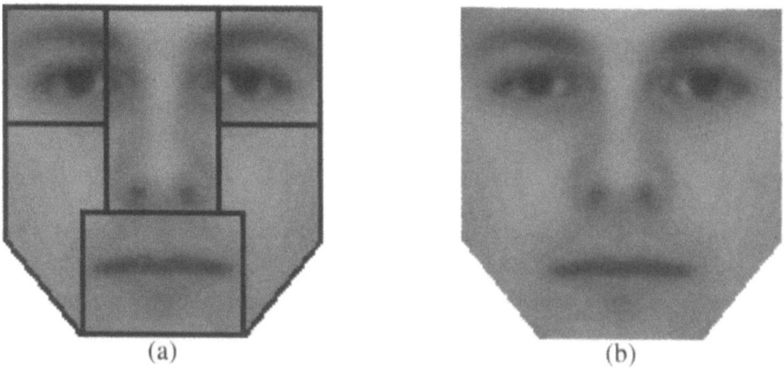

Figure 7.4: (a) The four facial regions (eye, nose, mouth, cheek) for which separate PCA bases are computed. (b) The average image computed over each of the regions independently using the 320 training images (160 ×2 using symmetry).

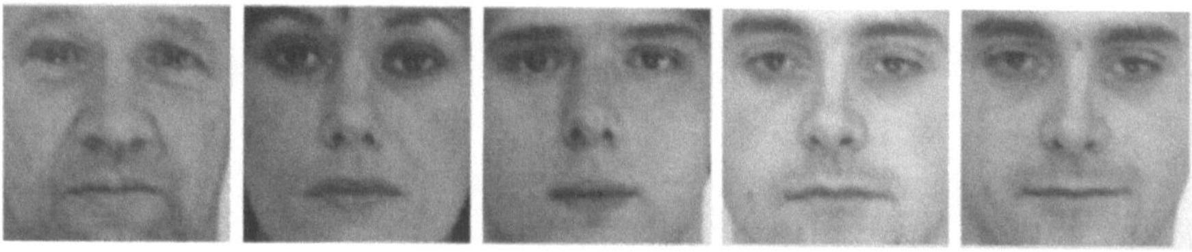

Figure 7.5: Five of the 160 training images. The faces have been registered using a similarity transformation computed from hand-placed marker points on the eyes and mouth.

are shown in Figure 7.6. The composed average regions are shown in Figure 7.4(b).

A face image is thus represented here by six sets of principal component coefficients (one for each of the left and right eyes, nose, mouth, left and right cheeks),

$$\begin{aligned} \mathbf{f} &= [\mathtt{V}_{\text{lefteye}}\mathtt{V}_{\text{righteye}} \dots \mathtt{V}_{\text{rightcheek}}]\mathbf{y} + \boldsymbol{\mu} \\ &= \mathtt{V}\mathbf{y} + \boldsymbol{\mu} \end{aligned} \tag{7.1}$$

where the matrices $\mathtt{V}$ are the projection matrices whose columns form the PCA basis for each feature, $\mathbf{y}$ is the stacked vector of coefficients associated with each basis, and $\boldsymbol{\mu}$ is a vector containing the average image features for each region. It is important to note that, by representing the face in terms of the sub-space parameterization $\mathbf{y}$, the dimensionality of the (in this case) 120×120 pixel images is reduced from 14 400 parameters down to just $319 \times 6 = 1914$ parameters. This reaps benefits in terms of both improved problem conditioning and faster optimization.

When using the PCA model, we often do not want to use *all* 319 basis vectors per feature. For instance, we would intuitively expect that the cheek, which is a smooth, low-detail feature, will required fewer principal components to accurately represent it than (say) the eye. Also, by reducing the number of components used to represent each feature to the lowest acceptable number, we are reducing the dimensionality of the model still further, and we would expect this to improve the condition of the super-resolution estimators. The criterion we will use to choose an appropriate di-

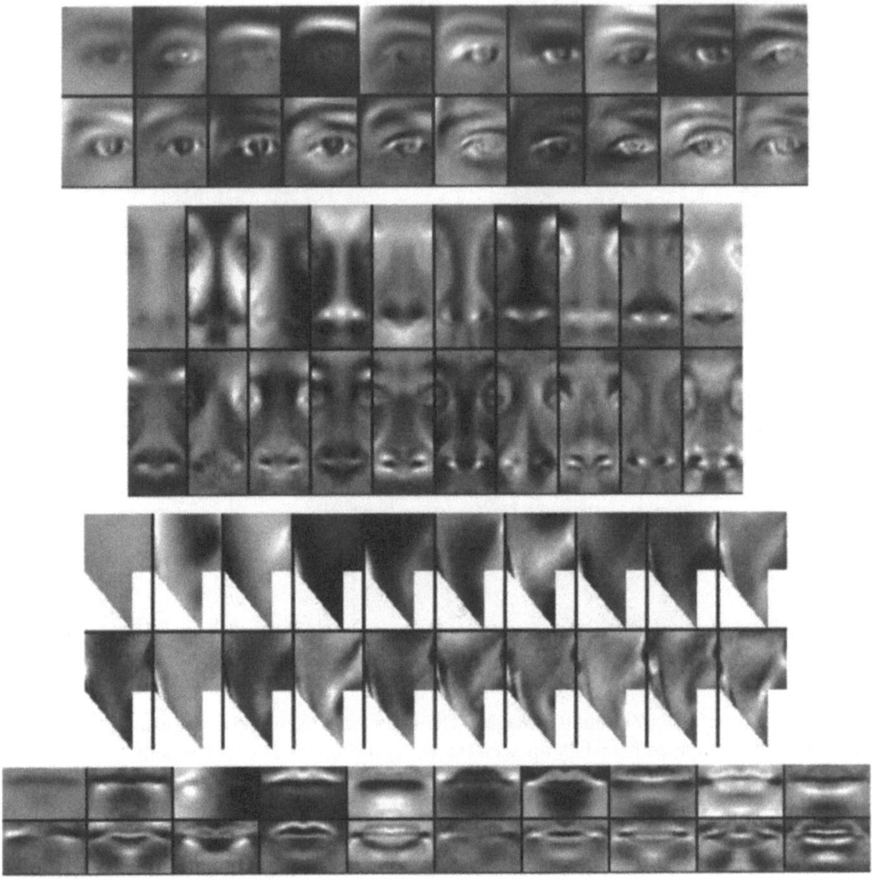

Figure 7.6: The first 20 eigenimages of each of the four facial features.

ν	0.95	0.98	0.99	0.995	1.0
Eye	27	64	103	149	319
Nose	39	92	142	190	319
Mouth	22	59	105	157	319
Cheek	8	29	69	126	319

Table 7.1: The number of principal components required to represent a particular fraction of the total variance within each feature.

mensionality for each feature is to keep the minimum number of components that span some fraction ν of the total variance. This calculation is extremely simple, since the variances of the principal components are obtained as a by-product of the PCA. Table 7.1 shows the required dimensionality of each feature in order to span various fractions of the total variance.

To demonstrate the ability of the model to represent a new face, Figure 7.7 shows

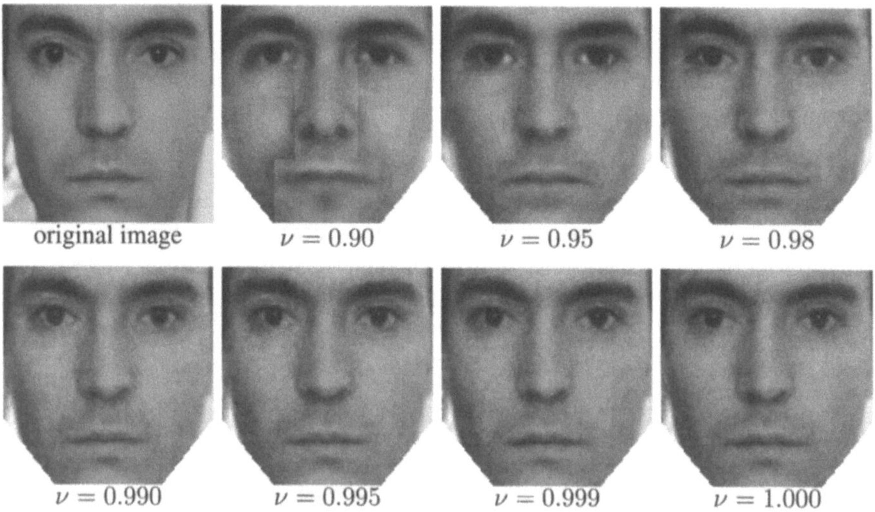

Figure 7.7: A face image *not* in the training set is projected onto the PCA face model. The number of components used to represent each feature is chosen to span some fraction (shown underneath) of the total variance.

a face which was *not in the training set* projected onto the PCA model, using feature dimensions chosen to preserve increasing fractions of the total variance. Reconstructions for which $\nu \geq 0.99$ are very close to the ground-truth.

7.4 Super-resolution using the PCA model

In this section we derive an ML estimator and two different MAP estimators based on the learnt face model.

7.4.1 An ML estimator (FS-ML)

The simplest way to use the PCA model is to constrain the super-resolution reconstruction to lie in the sub-space defined by the column span of $\mathtt{V}$. The ML solution using the sub-space parameterization follows directly from equation (5.26):

$$\mathbf{y}_{\mathrm{mle}} = \arg\min_{\mathbf{y}} \|\mathtt{M}(\mathtt{V}\mathbf{y} + \boldsymbol{\mu}) - \mathbf{g}\|^2 \tag{7.2}$$

for which the minimizer is given by

$$\mathtt{V}^{\top}\mathtt{M}^{\top}\mathtt{M}\mathtt{V}\mathbf{y} = \mathtt{V}^{\top}\mathtt{M}^{\top}(\mathbf{g} - \mathtt{M}\boldsymbol{\mu}) \tag{7.3}$$

Note that, as always, the average image is used to provide an estimate of super-resolution pixels *outside* the boundary of the reconstruction. In the case of the face model, this includes the missing triangular segments either side of the mouth. Also note that the number of parameters to be estimated is equal to the dimension of $\mathbf{y}$, the total number of principal components used.

7.4.2 MAP estimators

There are two straightforward ways to develop priors based on the face model, which are now described.

A prior over face-space (FS-MAP) A by-product of the PCA is an estimate of the variance of each principal component. This immediately gives us a very simple prior defined over the coefficients of the principal components. We use this to augment the ML estimator above, producing a MAP estimator :

$$\mathbf{y}_{\mathrm{map}} = \arg\min_{\mathbf{y}} \|\mathtt{M}(\mathtt{V}\mathbf{y} + \boldsymbol{\mu}) - \mathbf{g}\|^2 + \lambda \mathbf{y}^\top \Sigma^{-1} \mathbf{y}$$

for which the minimizer is given by

$$(\mathtt{V}^\top \mathtt{M}^\top \mathtt{M}\mathtt{V} + \lambda \Sigma^{-1})\mathbf{y} = \mathtt{V}^\top \mathtt{M}^\top (\mathbf{g} - \mathtt{M}\boldsymbol{\mu}) \tag{7.4}$$

where Σ is the diagonal matrix of component variances obtained from the PCA. We refer to this as a prior over face-space since, as with the ML estimator, the solution is constrained to lie on the sub-space defined by the PCA model, and the prior is defined in the sub-space. This estimator imposes a penalty proportional to the Mahalanobis distance from the features in $\mathbf{y}$ to the average features in $\boldsymbol{\mu}$. As with the FS-ML estimator, the number of parameters to be estimated is equal to the number of principal components that are used.

A prior over image-space (IS-MAP) A different way to use the learnt model is as part of a prior which encourages the estimated image to lie near to the PCA sub-space. We assume that the probability of obtaining $\hat{\mathbf{f}}$ is Gaussian in the distance of $\hat{\mathbf{f}}$ from $\mathtt{V}$. The resulting MAP estimator has the form

$$\mathbf{f}_{\mathrm{map}} = \arg\min_{\mathbf{f}} \|\mathtt{M}\mathbf{f} - \mathbf{g}\|^2 + \lambda \|(\mathtt{I} - \mathtt{V}\mathtt{V}^\top)(\mathbf{f} - \boldsymbol{\mu})\|^2$$

The minimizer is given by

$$(\mathtt{M}^\top \mathtt{M} + \lambda(\mathtt{I} - \mathtt{V}\mathtt{V}^\top))\mathbf{f} = \mathtt{M}^\top \mathbf{g} + \lambda(\mathtt{I} - \mathtt{V}\mathtt{V}^\top)\boldsymbol{\mu} \tag{7.5}$$

We refer to this as a prior over image-space, since the Gaussian distribution attached to the face-space defines a prior over all images $\mathbf{f}$. The solution is not constrained to lie in face-space.

For all three proposed estimators, the optimization is efficiently performed using conjugate gradient descent. Note however that the ML and FS-MAP estimators require optimization over the compact face-model parameters $\mathbf{y}$, and can therefore be computed much more rapidly than the IS-MAP estimator, which is parameterized in terms of the actual super-resolution pixels $\mathbf{f}$.

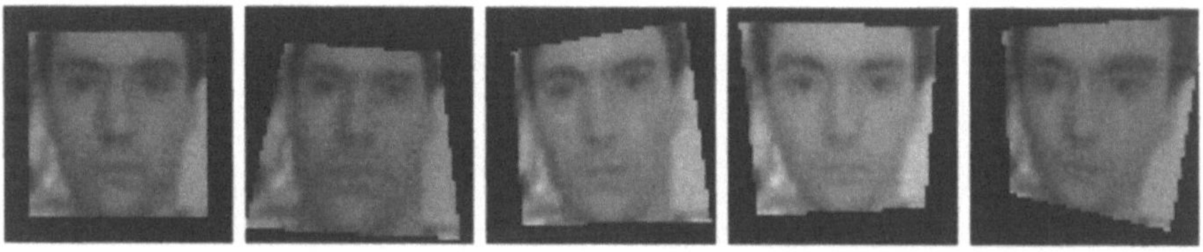

Figure 7.8: Five of the set of 30 synthetic images used to probe the behaviour of the face-based super-resolution estimators. The synthesis parameters are described in the text.

7.5 The behaviour of the face model estimators

To probe the behaviour of the three estimators, we use a sequence of 30 synthetic images, generated using the "Face" ground-truth image, where the PSF is $\sigma_{\text{psf}} = 0.7$ and the down-sampling ratio is $S = 3$. Gaussian noise with $\sigma = 5$ grey-levels is added to the images, five of which are shown in Figure 7.8. All of the super-resolution reconstructions are performed at $3\times$ zoom.

For the purposes of comparison, Figure 7.9 shows the ground-truth image, one of the low-resolution input images, the average image formed from all 30, and reconstructions at $3\times$ zoom using the three different MAP estimators from Chapter 6.

The ML estimator Figure 7.10 shows the constrained estimate computed at $3\times$ zoom with increasing numbers of principal components used per feature (according to variance fraction ν). In each case, all 30 low-resolution images are used. Initially, as ν increases, the reconstruction improves, moving away from the average face $\boldsymbol{\mu}$. But for $\nu > 0.99$, the reconstruction quality decreases rapidly as the more spurious and unstructured components allow noise to be introduced into the solution.

Figure 7.11 compares the face-space ML estimate as the number of input images is varied. In each case, the number of components used to represent each feature is chosen according to $\nu = 0.99$. As we might expect, an increasing number of images results in a reduced reconstruction error.

Note that in both of the above examples, the amount of noise on the observations ($\sigma = 5$ grey-levels) is far beyond the level that the regular ML estimator of Chapter 5 could possibly deal with.

Also note that, even in the noise-free case, we would *not* expect the face-space ML estimate to correspond exactly to the closest orthogonal projection of the ground-truth onto face-space (which is shown in Figure 7.7). The reason for this is as follows. The unconstrained ML framework of Chapter 5 effectively defines a multi-dimensional normal distribution over the space of all images $\mathbf{f}$. The unconstrained MLE, which is equal to the ground-truth in the noise-free case, sits at the mean of this distribution, and the covariance is determined by the matrix $\mathtt{M}^\top\mathtt{M}^{-1}$ as shown in equation (5.30). The ML estimate constrained to lie in face-space sits at the maximum of the distribution defined by the intersection of face-space with the normal distribution. This is equivalent to saying that it is the closest point in face-space to the unconstrained MLE (and hence to the ground-truth in the noise-free case) in terms of *Mahalanobis distance*. The situation is illustrated in Figure 7.12. In general, this will not be the same

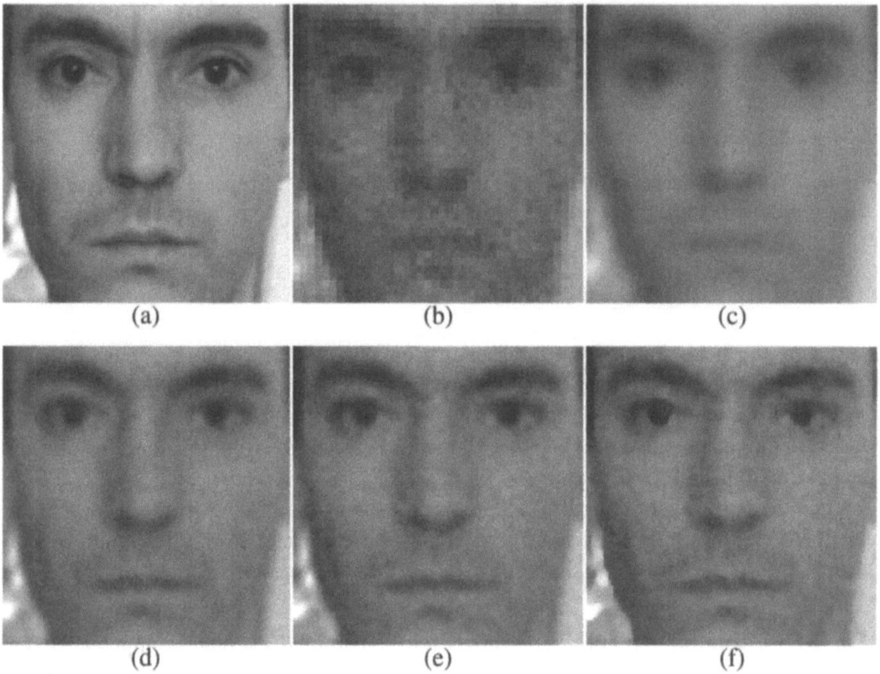

Figure 7.9: (a) The ground-truth "Face" image. (b) One of the 30 synthetic, low-resolution images. Reconstructions at $3\times$ zoom : (c) the average image, (d) MAP estimate with the simple Tikhonov $\|x\|^2$ prior ($\lambda = 0.05$), (e) using the first-derivative GMRF prior ($\lambda = 0.05$), (f) using the first-derivative HMRF prior ($\lambda = 0.05, \alpha = 0.025$).

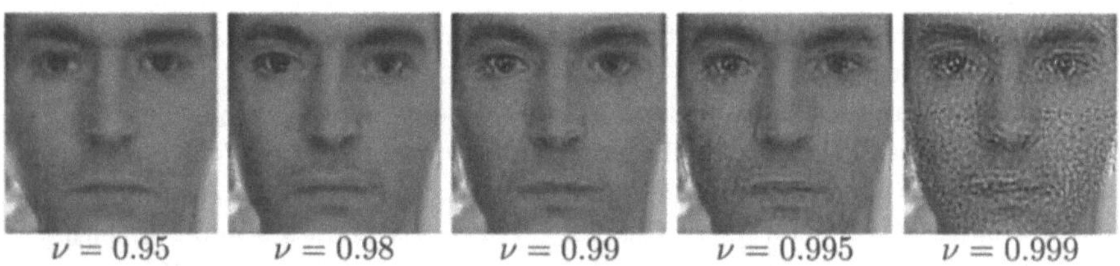

Figure 7.10: The face-space constrained ML estimate computed from 30 images at $3\times$ zoom, using an increasing number of principal components to represent each facial feature (chosen according to ν). As ν increases, the reconstruction moves away from the average face $\boldsymbol{\mu}$, and the accuracy improves. But for $\nu > 0.99$ the final few components, which are typically less structured, allow excessive noise into the solution.

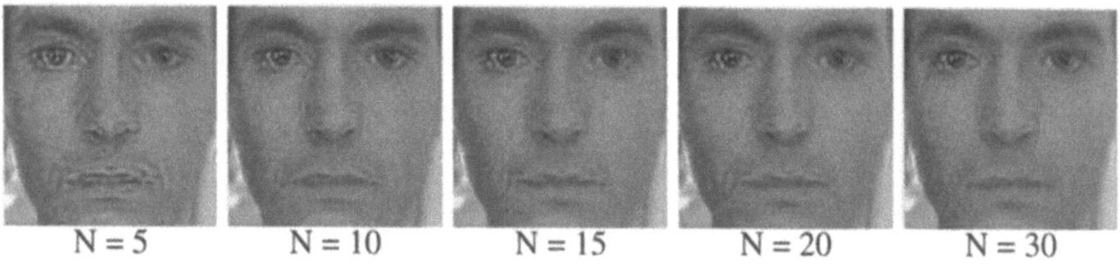

Figure 7.11: The face-space ML estimate computed using an increasing number of low-resolution images, with the number of components per feature fixed according to $\nu = 0.99$. As we would expect, the reconstruction improves as the number of images increases.

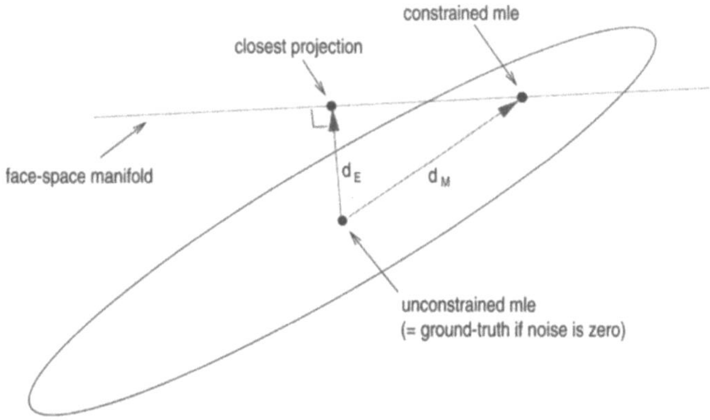

Figure 7.12: The face-space constrained MLE is the closest point in face-space to the unconstrained MLE in terms of Mahalanobis distance d_{M} (defined by the normal equations (5.27)). This can be very different from the closest reprojection in terms of Euclidean distance d_{E}. Hence, even in the noise-free case, when the unconstrained MLE is equal to the ground-truth, the constrained MLE is does not generally correspond to the orthogonal projection of the ground-truth onto face-space.

as the orthogonal reprojection, which is the closest face-space point to the ground-truth in terms of *Euclidean distance*. Because the matrix $\mathtt{M}^\top\mathtt{M}$ is very ill-conditioned, the normal distribution will be severely elongated in certain directions. Consequently, even in the noise-free case, the constrained MLE can lie some distance from the unconstrained MLE.

MAP with face-space prior (FS-MAP) When using the MAP estimator defined over face-space we can afford to use all of the principal components, since the prior ensures that those which contribute only a tiny amount of variance will be heavily surpressed in the super-resolution reconstruction.

Figure 7.13 compares the MAP estimate as the value of the prior influence varies. Even though λ varies by a factor of 20 in these examples, the estimate remains fairly constant, with just a small amount of reconstruction error entering as λ becomes very small. This is probably due to the fact that the variance of the principal components falls off very rapidly. For instance, in the case of the mouth components, the variance

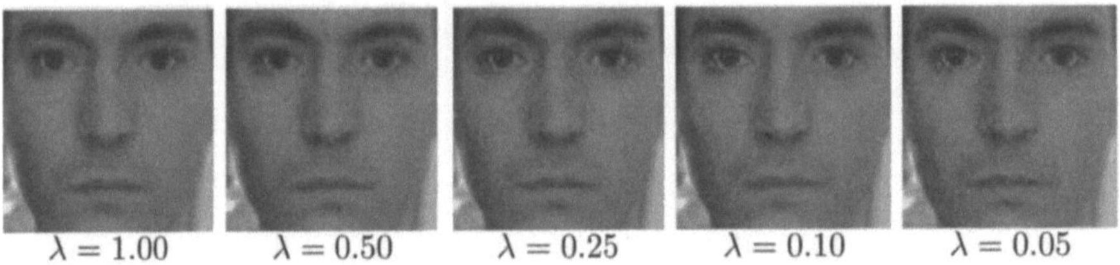

Figure 7.13: Reconstructions using the FS-MAP esimator. As λ varies by a factor of 20, the reconstruction remains fairly constant. At very high values of λ the solution tends toward the average face $\boldsymbol{\mu}$. When λ is zero, the solution is equivalent to the constrained FS-ML estimator.

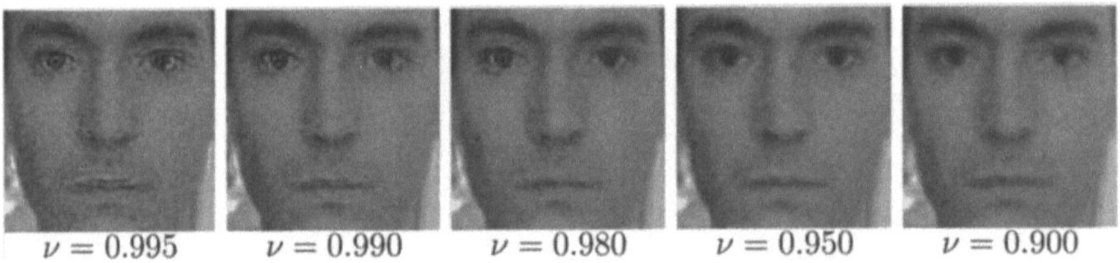

Figure 7.14: Reconstructions using the IS-MAP estimator for varying numbers of principal components per feature. λ is fixed at 0.1. The optimal number of components per feature in this example appears at around $\nu = 0.98$.

of the 40th component is 1000 times smaller than that of the first component. Consequently, even a small contribution from this prior is enough to severely dampen all but the first few tens of principal components. For very high values of λ, $\mathbf{y}$ tends toward zero, and hence the super-resolution estimate tends toward the average face $\boldsymbol{\mu}$. When λ is zero, the solution is equivalent to the constrained FS-ML estimator.

MAP with image-space prior (IS-MAP) For the MAP estimator based on the image-space prior, both the number of components per feature and the prior influence λ. Figure 7.14 compares the MAP estimate as ν varies with λ fixed at 0.1. If ν is too large, the solution tends to become noisy, much like the FS-ML estimate. If ν is too small, the sub-space is not descriptive enough; the boundaries between the facial features start to become apparent, and the quality of the estimate decreases. The optimum value in this example is around $\nu = 0.98$.

Figure 7.15 shows the effect of varying λ whilst fixing ν at 0.9. If λ is zero, the estimate will be identical to the unconstrained MLE. At the other extreme, when λ is large, the result is almost equivalent to the constrained ML estimate as the image is forced to lie on the face sub-space. Between the two, the solution has some freedom to lie a short distance from the sub-space, and this can ameliorate some errors due to inadequacies of the face model. The benefit can be seen in the closeness to the ground truth at around $\lambda = 0.1$.

Finally, Figure 7.16 compares the HMRF-MAP result from Figure 7.9 with the best results using the face-specific MAP estimators. Arguably, both FS-MAP and IS-MAP estimators give a better results than the HMRF, although the FS-MAP estimate

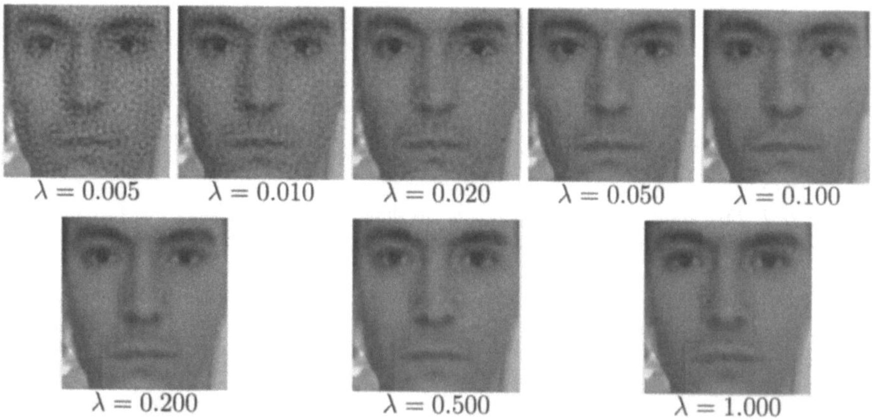

Figure 7.15: IS-MAP reconstructions as λ varies, with ν fixed at 0.9. For small λ the solution will approach the unconstrained MLE. For large λ the solution approaches the face-space constrained MLE.

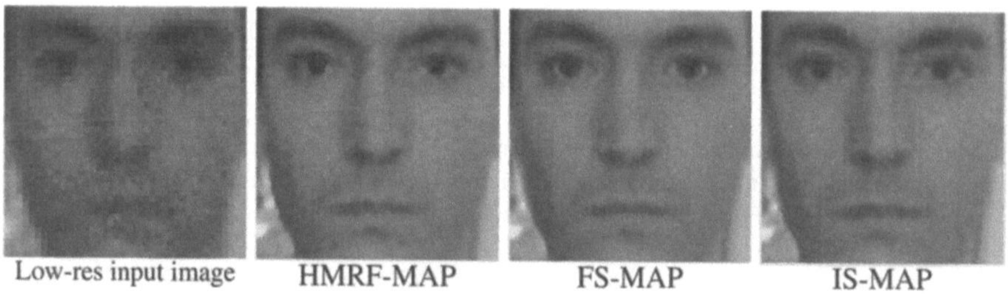

Figure 7.16: Comparison of the HMRF-MAP estimate of Figure 7.9 with the best results of the two face-specific MAP estimators. The FS-MAP estimate seems to be the best, reconstructing both smooth and detailed areas fairly accurately. All three are of vastly superior quality to the input images, one of which is shown.

appears to be the best, providing a fairly good reconstruction of both smooth and detailed areas. All three are of vastly superior quality to input images, one of which is shown.

7.6 Examples using real images

Fixed camera, moving face Figure 7.17 shows five frames from a sequence of 25 of a moving face[1] captured using a monochrome Cohu CCD camera. The faces were initially registered using the feature-based N-view maximum-likelihood method, and the photometric parameters were initialized using the method of Chapter 3. Having obtained this initial registration, the average image was formed. Finally, the registration and photometric parameters were refined by optimizing the super-resolution

[1]Coincidentally, this face and the one used in the synthetic examples belong to the same person. Experimental volunteers are somewhat scarce in this very dangerous area of research.

Figure 7.17: Five frames from a sequence of 50 showing a moving face. The face occupies 40×40 pixels in these images.

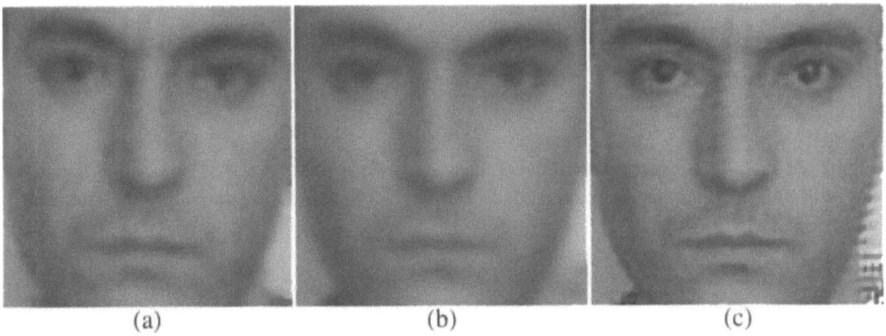

Figure 7.18: For the sequence shown in Figure 7.17: (a) the region of interest in one of the input images (shown at $3\times$ zoom using bicubic interpolation); (b) the average image; (c) the super-resolution HMRF-MAP estimate at $3\times$ zoom with $\lambda = 0.025, \alpha = 0.05$.

likelihood function of equation (5.25) with respect to the parameters $(\mathtt{H}, \alpha, \beta)_n$ of each image, keeping $\mathbf{f} = \mathbf{f}_{\text{avg}}$ fixed. The face occupies a 40×40 pixel region. The super-resolution reconstructions are 120×120 pixels, i.e. $3\times$ pixel zoom.

For the purpose of later comparison, Figure 7.18 shows the region of interest in one of the input images; the average image; and the super-resolution reconstruction using the HMRF-MAP estimator with $\lambda = 0.025, \alpha = 0.05$ (tuned by trial-and-error).

Figure 7.19 shows reconstructions using the face-space constrained ML estimator as the number of components per features varies. The FS-ML estimator performs surprisingly well on this data set. The reconstruction for which $\nu = 0.995$ shows good detail with relatively little noise.

Figure 7.20 shows reconstructions using the FS-MAP estimator as λ varies. As with the synthetic examples, the quality of reconstruction is consistently quite high, even though λ varies over a wide range.

Figure 7.21 shows the reconstructions using the IS-MAP estimator as λ varies, ν being fixed as 0.99. When $\lambda = 0.005$, the estimate tends toward the unconstrained MLE, and reconstruction error is evident. But when $\lambda = 0.05$, the reconstruction is very good.

Finally, Figure 7.22 compares the quality of the input images, the HMRF-MAP estimate and the IS-MAP estimate with $\lambda = 0.05, \nu = 0.99$. Both reconstructions are very good, but the IS-MAP estimator produces better detail around the eyes, nose and

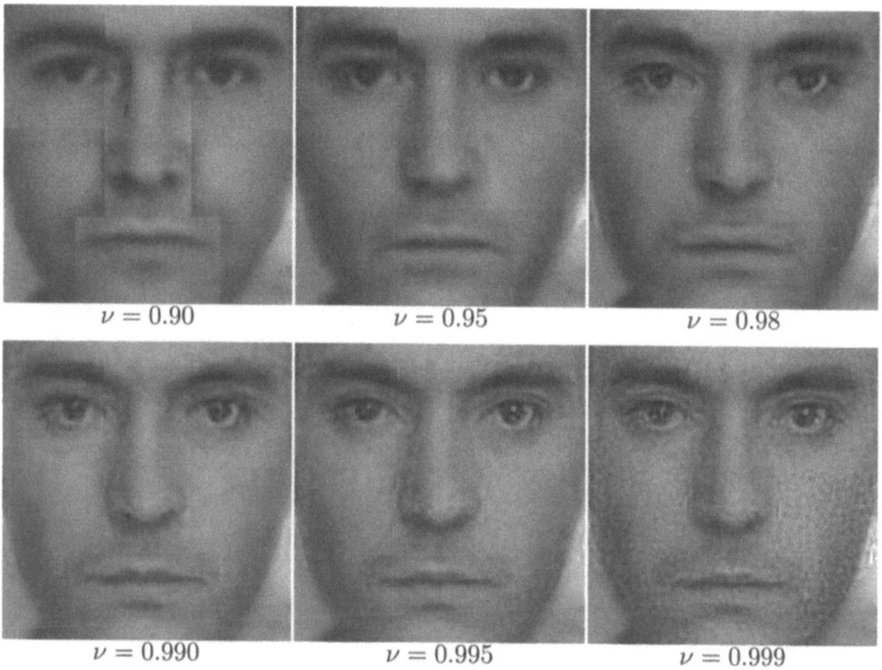

Figure 7.19: FS-ML estimates at $3\times$ zoom for varying numbers of components per feature. As ν increases the estimate moves away from the average face $\boldsymbol{\mu}$.

mouth.

Fixed face, moving camera Figure 7.23 shows a mosaic featuring a face. The mosaic was created from a sequence of 30 PAL-size, JPEG compressed images captured using a Cohu CCD camera which was rotated on a tripod. Eight of the original frames are also shown. The face occupies only 30×30 pixels in the low-resolution images.

Figure 7.24 shows reconstructions using the FS-MAP and IS-MAP face-space estimators with various parameters settings. For comparison, one of the input images, the average image, and a reconstruction using the HMRF-MAP estimator are also shown. The size of the reconstructed image is 120×120 pixels, 16 times as many pixels as in the 30×30 pixel low-resolution region of interest. The $4\times$ zoom ratio and the fairly poor quality of the input imagery means that the face-space ML estimator does not produce good results and is omitted here. In this example, the quality of the IS-MAP estimates is arguably equal or superior to the HMRF estimate, although some artifacts are visible on the boundaries between the different facial features.

FS-MAP vs. IS-MAP In contrast to the results obtained using synthetic sequences, in the real image examples IS-MAP gives slightly better results than FS-MAP. It is not entirely clear why this should be the case, although it may indicate that IS-MAP is more robust to registration error and/or deviations from the Gaussian noise model than

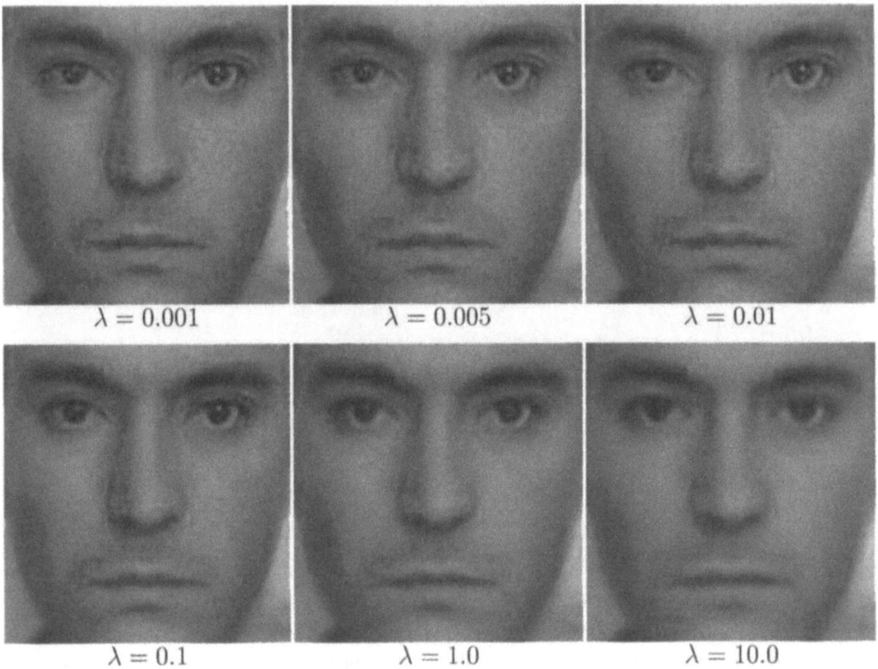

Figure 7.20: FS-MAP reconstructions as λ varies. The quality of the reconstruction is consistently high over a large range of values of λ.

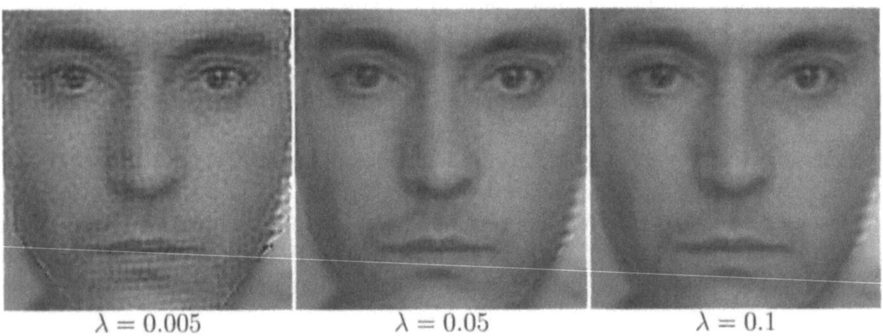

Figure 7.21: IS-MAP reconstructions as λ varies, ν being fixed at 0.99.

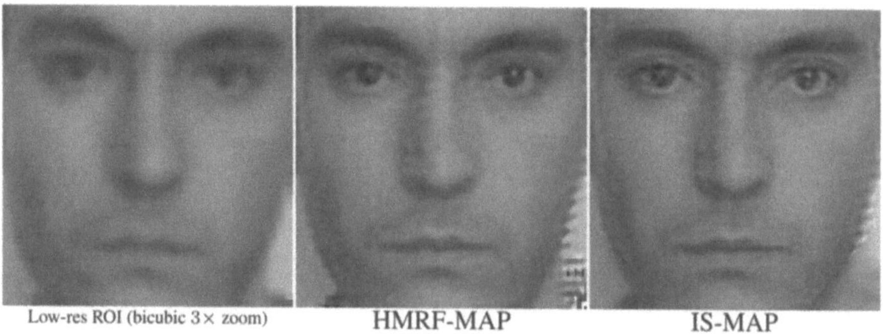

Figure 7.22: A comparison of the quality of the input imagery with the HMRF-MAP estimate, and the face-model IS-MAP, all at $3\times$ zoom from 25 images. The IS-MAP estimator produces a more veridical result around the eyes, nose and mouth.

FS-MAP, since neither of these effects were investigated in the synthetic examples. However, to fully resolve this issue would require a much larger set of examples.

Choosing λ In these examples, λ has been systematically varied in order to demonstrate its effect. In practice however, the same cross-validation method as described in Section 6.7 may be used to choose an optimal value of λ.

Figure 7.23: (Top) Eight of 30 PAL-size, JPEG compressed images captured using a rotating CCD camera. (Bottom) A mosaic created from the image sequence. Figure 7.24 shows the results of super-resolution enhancement applied to the face.

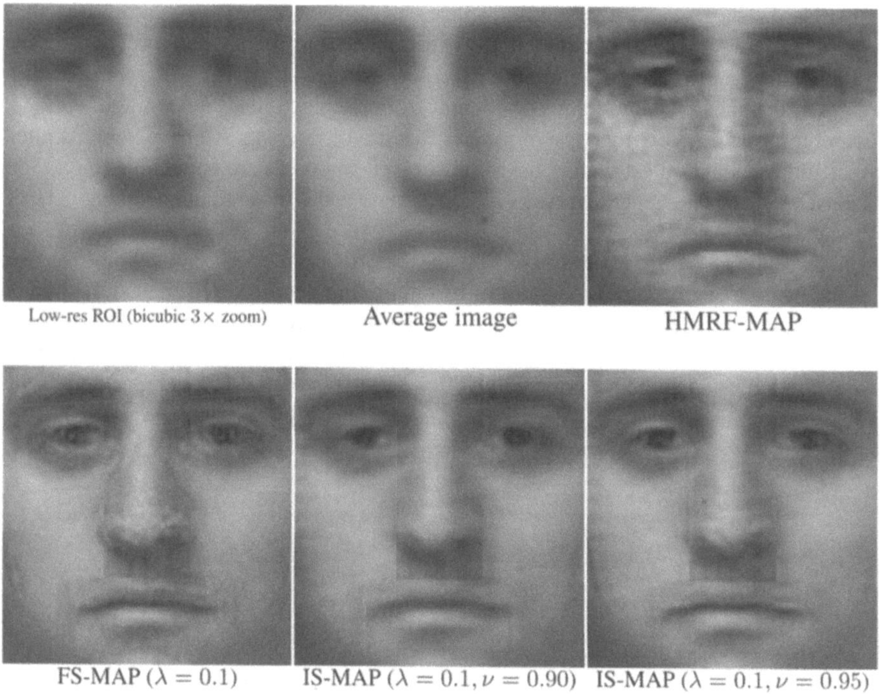

Figure 7.24: Super-resolution estimation from 30 images. (Top) The region of interest in one of the low-resolution input images; the average image; and the HMRF-MAP estimate with $\lambda = 0.025, \alpha = 0.075$. (Bottom) Super-resolution reconstructions using the FS-MAP and IS-MAP face-model priors method. The size of the reconstructed image is 120×120 pixels, 16 times as many pixels as in the 30×30 pixel low-resolution region of interest. The quality of the IS-MAP estimates is arguably equal or superior to the HMRF estimate, although some artifacts are visible on the boundaries between the different facial features.

7.7 Summary

In this chapter we have described the use of problem-specific sub-space models for the super-resolution reconstruction of two different types of image. The use of sub-space models has several advantages over the generic priors seen in the previous chapter. First, they can regularize the super-resolution inverse problem without imposing undesirable spatial correlations between pixels in the solution. Second, in certain cases, such as the face model, the solution space can be represented very compactly, thus reducing the number of parameters to be estimated and dramatically decreasing the computational effort required to obtain a solution. Third, in certain cases, the parameters of the sub-space model may be learnt from training data, thus tuning the model to represent accurately the type of images that are of interest. We have demonstrated that super-resolution estimators based on sub-space models are applicable to real image sequences, and generate results which at least equal those obtained using generic image priors.

The quality of the results based on the face-space model may be improved by using a larger set of training images, or by employing a more sophisticated method of registering the face images, such as one based on a deformable mesh. The artifacts which are sometimes visible on the boundaries between the different features may be ameliorated by including a prior term which penalizes spatial incoherence across the boundaries.

Chapter 8

Conclusions and Extensions

8.1 Summary

This book has investigated the problem of combining information contained in multiple, overlapping views of a scene into a single, high-resolution still image. There were three main threads to this investigation: image registration, image mosaicing, and super-resolution.

Image registration A robust and efficient algorithm has been described for finding corresponding feature points in two images and for obtaining an estimate of the homography relating them using the method of maximum likelihood. The sub-pixel accuracy of the method has been demonstrated empirically, and the statistical assumptions regarding the localization of the Harris feature points have also been verified experimentally.

A simple scheme for modelling the global photometric differences between images has been described, along with a robust method for estimating the parameters of this transformation given geometrically registered views. The model has proved effective in compensating for colour-space transformations due to various effects including camera gain and white balance.

Finally, the effectiveness of these registration algorithms has been demonstrated in a change detection application which allows latent marks to be extracted from confusing, non-periodic backgrounds. The method produces very good results when applied to real forensic images: for the separation of finger-marks from scanned bank notes; and foot-marks from digital photographs of floor tiles.

Image mosaicing It has been shown how the ML estimator of the homography relating a pair of views can be generalized to simultaneously estimate the set of homographies mapping multiple, overlapping views into a common frame. To enable this computation, a novel algorithm has been described for the efficient matching of corresponding interest points across multiple views. The resulting N-view ML estimator has been demonstrated to be immune to the "dead-reckoning" error accumulation which is possible when concatenating pair-wise image transformations over long sequences.

The photometric registration algorithm has proved effective for the removal of "seams" in image mosaics caused by variation in automatic gain control and white balance over the course of a sequence.

Based on these methods, mosaic images have been composed from a variety of different image sequences, the longest of which combines 177 images into a full 360° panorama.

Super-resolution The implementation of a generative image model for use in super-resolution reconstruction has been discussed in detail and an efficient implementation described. An intuitive picture of the behaviour of the ML super-resolution estimator has been developed. Using synthetic ground-truth data, the effects of image noise, registration inaccuracy, and variation in point-spread function have been investigated. By appeal to both analytical and empirical results, the dependency of the reconstruction error on the input image noise, point-spread function, number of images used, and zoom ratio has also been examined. The classic error back-projection algorithm of Irani and Peleg [84, 85] has been analysed and its relationship to the ML estimator highlighted.

The Bayesian framework for super-resolution restoration has been described, and the general form of the super-resolution *maximum a posterior* (MAP) estimator derived. The relationship between the Wiener filter and certain MAP estimators has been examined, showing how certain spatial priors may be interpreted in the frequency domain. Several "generic" Markov random field (MRF) prior models have been described and the characteristics of the corresponding MAP estimators investigated. A novel algorithm has been described which uses cross-validation to set the level of influence that the prior model has on the MAP solution in order to minimize reconstruction error without introducing excessive smoothness. The algorithm has been shown to produce near-optimal results on both real and synthetic examples.

A novel approach to super-resolution restoration based on problem-dependent sub-space models has been proposed. The method has been applied to two different classes of image: text and faces. In the latter case, the sub-space model is learnt directly from training images. In the case of text images, a simple constrained ML estimator has been shown to perform as well as the best MRF-MAP estimator, but without the need to introduce spatial correlations between pixels. In the case of face images, the learnt sub-space model has been used to define an ML estimator and two different MAP estimators. The sub-space estimators have been shown to perform at least as well and frequently better than the MRF-MAP estimators.

All of the super-resolution algorithms described have been applied very successfully to a variety of real, uncalibrated image sequences.

8.2 Extensions

This section discusses some possible avenues for future super-resolution research.

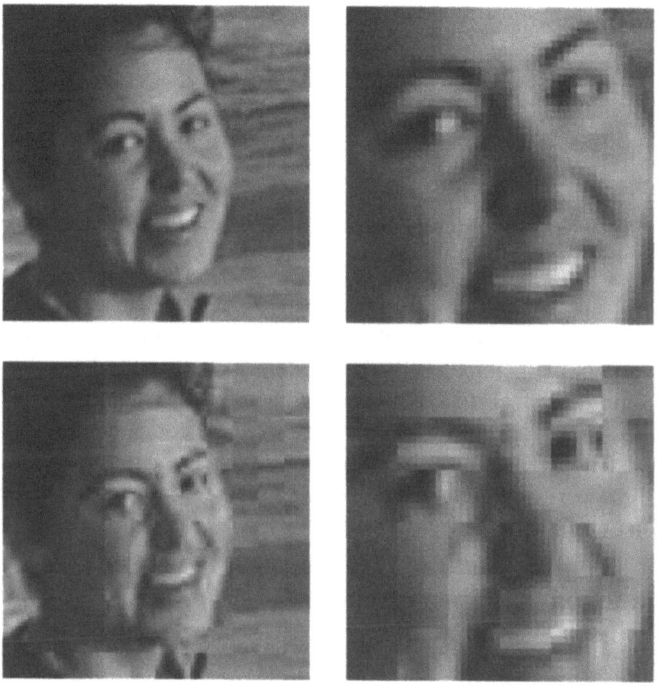

Figure 8.1: (Top) An uncompressed image. (Bottom) The image after JPEG compression. Artifacts are clearly visible in the close-up of the compressed image due to excessive loss of high-frequency information.

8.2.1 Application to digital video

Modern video imagery is, almost without exception, subject to some form of lossy compression. The current crop of compression methods, such as Motion-JPEG and MPEG, are based on block-wise transform encoding. The image is divided into typically 8×8 pixel blocks and each block is projected into an orthogonal basis, such as the *discrete cosine transform* (DCT). Transform components with coefficients below a threshold are dropped, and the remaining coefficients are quantized, packed into space-efficient data structures, and finally compressed using a loss-less method. Decompressing/reconstructing the image is simply the inverse process, but the information lost in discarding and quantizing the original transform coefficients means that the decompressed image will not be identical to the original. Globally, the difference is hardly noticeable, but at the level of the individual blocks, the difference can be severe. The problem is illustrated in Figure 8.1, which shows an image before and after JPEG compression. Although globally the images appear very similar, the close-ups reveal that high-frequency information has been lost.

Consequently, a major source of degradation in modern video images is due to lossy compression. However, the compression/decompression operations could be thought of as providing an explicit generative model describing the transformation of the original high-resolution image into the degraded, decompressed image. Unlike

Figure 8.2: Due to blur, many pixels in the degraded image depend on two or more characters in the number plate.

the generative model used throughout this book , whose parameters involve the blur and sampling rate, the parameters of this modern model would depend on knowing which transform components were dropped in each block, and on the level of quantization applied. These values may be recovered from the compressed image file itself. Such a generative model could be used as a direct replacement for the one described here, allowing super-resolution restoration from multiple transform-encoded images. Inspiration in this direction may be provided by [72, 133, 162].

8.2.2 Model-based super-resolution

In Chapter 7 we investigated model-based super-resolution algorithms, in which a model specified for a particular class of image is used either to constrain the super-resolution estimate, or as the basis for a Bayesian prior. The potential of this approach was demonstrated for images of faces, for which a simple, very specific model can be easily developed. Here we discuss two further areas of application.

Vehicle registration plates Another potential area of application is in the restoration of vehicle registration plates from degraded surveillance video. Registration plates conform to a very restricted set of possible formats and font types. In fact, as of August 2001, all new UK registration plates must legally conform to an exactly specified format and font. Consequently, a very low-dimensional model has already been provided, the parameters being simply the seven characters on the plate, which consists of five letters and two numbers. There are therefore around $26^5 + 10^3$ total configurations (ignoring the fact that some combinations are not allowed).

The problem with using such a model is in the optimization stage. Since the parameter space is discrete, and hence not differentiable, gradient-based optimization techniques cannot be employed. The size of the configuration space is far to large to permit an exhaustive search. If the observed pixels could be partitioned into independent sets of observations, such that within each set the observations depend only on a single character, then each character could be optimized independently of the others, and a simple brute force search would suffice, whereby the observation likelihood is evaluated for each possible state.

Unfortunately, such a partitioning is not generally possible, since the effect of blur means that many low-resolution pixels are dependent on two or more adjacent characters. The problem is illustrated in Figure 8.2. Some form of stochastic global-optimization method must therefore be employed. This typically involves evaluation of the observation likelihood for a vast number of trial solutions, each evaluation requiring the plate to be projected through the generative model, which is a rather expensive operation. Consequently, naive application of simulated annealing or a genetic algorithm may prove prohibitively expensive due to the high-cost of evaluating putative solutions.

An effective solution to this problem would probably involve as a first step the selection of a small number of candidates for each character on the plate. This could be achieved by performing an approximate partitioning of the observations, followed by brute force likelihood evaluation over all states for each individual character or pair of adjacent characters. Having reduced the configuration space in this way, a stochastic method may be applied more successfully. Mean-field or belief propagation [171] techniques may also offer an efficient method of solution.

Learnt image models Part of the attractiveness of the model-based approach is the potential for learning image models from sets of training images. The recent work of Freeman and Pasztor [62, 64] may be a useful pointer in this direction. In their method, an image is represented by a uniform grid of overlapping blocks. Each block is drawn from a lexicon of exemplars, which are learnt by vector quantization of blocks sampled from the training images. The smaller the blocks, the more generic are the image features compiled into the lexicon. The size of the blocks must be small enough to allow a reasonable size lexicon to be compiled, but large enough to capture something meaningful about the small-scale structures present in the images. Finally, a compatibility constraint on the overlap between blocks encourages long-range order.

Freeman and Pasztor's application is the direct inference of missing high-frequency detail in single images. The basic principal is as follows. Training images are split into high- and low-frequency parts, and a block-wise lexicon is learnt for each frequency band. Given a new image, inference is performed by projecting the image block-wise onto the low-frequency lexicon, and then inferring the corresponding high-frequency blocks by cross-referencing the two lexicons. The cross-references are determing by the co-occurences of particular high–low frequency blocks in the training images.

The authors reason that the lowest spatial frequencies are of little use in inferring the missing high frequencies, and therefore that including the very lowest frequency components in their model constitutes an unnecessary modelling burden. Consequently, they only apply learning to mid-pass and high-pass filtered training images. This idea may also be exploited in the super-resolution framework, where an excellent approximation of the low-frequency part of the super-resolution image can be computed very cheaply using the average image, or a MAP estimator with a strong smoothness prior. Only a single lexicon would need to be learnt, capturing the high-frequency image structure. The compact image model would then consist of the fixed, precomputed low-frequency part together with the exemplar-based model of the high-frequency part.

The challenge in using such a model is similar to that posed by the registration plate model. Each block has a discrete parameter – an index into the potentially large lexicon of exemplars – and one or more continuous parameters specifying a uniform intensity transformation. Again, a feasible means of solution will require a method for selecting a handful of high-likelihood candidate exemplars for each block. A stochastic optimization strategy, or possibly a belief propagation method, may then be employed.

8.3 Final observations

The success of both the mosaicing and the super-resolution algorithms described here depends heavily upon the accuracy of the geometric registration. Registration inaccuracy in mosaicing causes inconsistencies due to sequential error accumulation over long sequences. In super-resolution, it leads to correlated errors in the reconstructed image. The fact that good results have been demonstrated in both areas is further evidence, if any were needed, that the ML homography estimation based on Harris point-features is indeed an excellent, accurate method.

Image mosaicing and super-resolution offer two different solutions to the problem of generating a high-resolution, wide field-of-view (FOV) image: mosaicing multiple, zoomed-in (narrow FOV) images; or super-resolution restoration applied to multiple, zoomed-out (wide FOV), low-resolution images. Given the computational expense involved in applying super-resolution restoration to large images, it is probably safe to conclude that "video paintbrush" mosaicing offers by far the most efficient solution. However, image mosaicing may easily be combined with an inexpensive, multiple image resampling method, such as that proposed by Rudin *et al.*, thus providing a very powerful tool.

One of the basic ideas motivating super-resolution research is driven by a counting argument: that by using many images, the number of observations is much greater than the number of parameters to be estimated, and hence the ill-posed single image restoration problem can be turned into a well-posed one. Indeed, providing the image motion is not degenerate, this argument holds true. But as we have seen, where super-resolution is concerned, *well-posed* does not necessarily equate with *well-conditioned.* In the current literature, it is always assumed that super-resolution implies an increase in pixel density (by at least a factor of 4), affording the super-resolution image the ability to represent frequencies beyond the Nyquist limit of the input images. As we have demonstrated, it is the pixel-density, or zoom ratio, that has the most dramatic effect on the condition of the restoration problem. It is perfectly possible to obtain acceptable results using the unconstrained ML estimator for modest zoom ratios. If blur or noise is the dominant degradation in the source imagery, then an increase in pixel density is not necessary for the restored image to have a much higher perceived resolution. By taking advantage of multiple images to perform deblurring and denoising only, text may be made legible, and faces recognizable. Furthermore, we have seen the potential for model-based maximum-likelihood methods to produce super-resolution results comparable to those determined by Bayesian methods. In summary, there is still a place for the careful use of maximum-likelihood methods in super-resolution, and awareness of this fact will hopefully benefit imaginative future researchers.

Appendix A

Large-scale Linear and Non-linear Optimization

This appendix outlines the efficient methods which are employed for the iterative solution of very large, sparse systems of equations, both linear and non-linear. To aid visualization, the methods are illustrated here using examples with only two variables. However, the methods may be used in problems featuring thousands of parameters.

A.1 Notation

The general problem is the minimization of a scalar function of N variables

$$\mathbf{x}_{\min} = \arg\min_{\mathbf{x}} f(\mathbf{x}) : \Re^n \rightarrow \Re$$

where $\mathbf{x}$ is an N-dimensional vector of parameters. The gradient $\nabla f(\mathbf{x})$ of $f(\mathbf{x})$ is the vector

$$\nabla f(\mathbf{x}) = \left[\frac{\partial f}{\partial x_1} \cdots \frac{\partial f}{\partial x_N} \right]$$

The Hessian $\mathtt{H}(\mathbf{x})$ of $f(\mathbf{x})$ is the symmetric matrix

$$\mathtt{H}(\mathbf{x}) = \begin{bmatrix} \frac{\partial^2 f}{\partial x_1^2} & \cdots & \frac{\partial^2 f}{\partial x_1 \partial x_N} \\ \vdots & \ddots & \\ \frac{\partial^2 f}{\partial x_1 \partial x_N} & & \frac{\partial^2 f}{\partial x_N^2} \end{bmatrix}$$

A.2 Quadratic functions

A multi-dimensional quadratic function has the form

$$f(\mathbf{x}) = a + \mathbf{b}^\top \mathbf{x} + \frac{1}{2} \mathbf{x}^\top \mathtt{C} \mathbf{x}$$

If $\mathtt{C}$ is *symmetric, posi-definite* then the curvature is everywhere positive, and the function has a unique minimum at

$$\mathbf{x}_{\min} = -\mathtt{C}^{-1}\mathbf{b}$$

When N is large, it is not feasible to perform this inversion directly. There are two common situations in which such systems arise :

Linear least squares

$$\begin{aligned}\mathbf{x}_{\min} &= \arg\min_{\mathbf{x}} \|\mathtt{A}\mathbf{x} - \mathbf{b}\|^2 \\ &= \arg\min_{\mathbf{x}} (\mathbf{b}^\top\mathbf{b} - 2\mathbf{b}^\top\mathtt{A}\mathbf{x} + \mathbf{x}^\top\mathtt{A}^\top\mathtt{A}\mathbf{x})\end{aligned}$$

Local Taylor series approximation

$$f(\mathbf{x}+\mathbf{h}) \approx f(\mathbf{x}) + \nabla f(\mathbf{x})^\top\mathbf{h} + \frac{1}{2}\mathbf{h}^\top\mathtt{H}(\mathbf{x})\mathbf{h}$$

A.2.1 Steepest descent

The simplest means of solution is the method of steepest descent, in which minimization of the N-dimensional function proceeds by a series of 1D line-minimizations

$$\mathbf{x}_{n+1} = \mathbf{x}_n + \lambda_n\mathbf{p}_n$$

The steepest descent method chooses $\mathbf{p}_n$ to be parallel to the gradient

$$\mathbf{p}_n = \nabla f(\mathbf{x}_n)$$

Step-size λ_n is chosen to minimize $f(\mathbf{x}_n + \lambda_n\mathbf{p}_n)$. For quadratic forms there is a closed form solution :

$$\lambda_n = \frac{\mathbf{p}_n^\top\mathbf{p}_n}{\mathbf{p}_n^\top\mathtt{C}\mathbf{p}_n}$$

Figure A.1 shows a quadratic function of two variables which has its minimum at $(5,5)$. Figure A.2 shows the steps taken by the steepest descent method in seeking the minimum. Note that, in a contour plot, the function gradient is everywhere perpendicular to the contour lines. After each line minimization the new gradient is *always* orthogonal to the previous step direction (this is true of any line minimization). Consequently, the iterates tend to zig-zag down the valley. In general, steepest descent is an *extremely slow and computationally inefficient* method of optimization.

A.2.2 Conjugate gradient descent

The method of conjugate gradient descent [69, 116] chooses successive descent directions $\mathbf{p}_n$ such that it is guaranteed to reach the minimum in a finite number of steps.

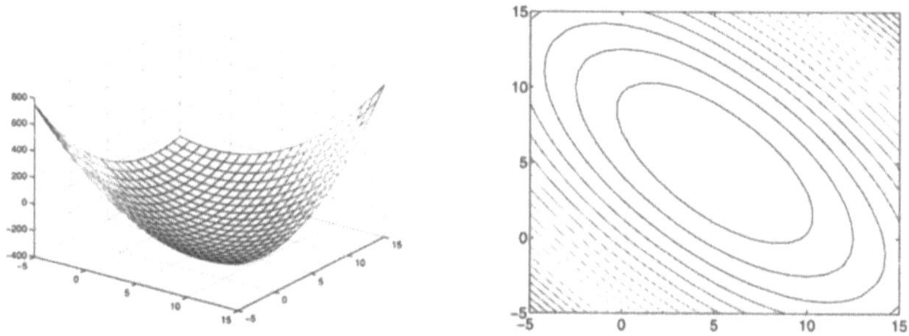

Figure A.1: A simple quadratic function of two variables.

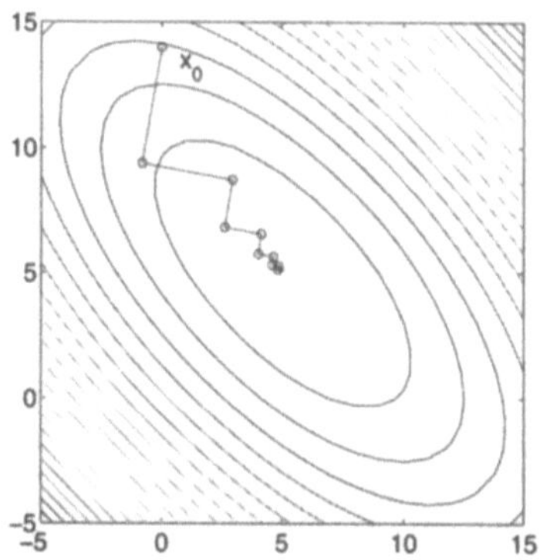

Figure A.2: The minimum of the quadratic function of Figure A.1 is sought by the method of steepest descent. The iterates zig-zag down the valley. Steepest descent is a slow and inefficient method of minimization.

Each $\mathbf{p}_n$ is chosen to be conjugate to all previous search directions with respect to the Hessian C,

$$\mathbf{p}_n^\top \mathtt{C} \mathbf{p}_j = 0, \quad 0 \leq j < n \tag{A.1}$$

The resulting search directions are mutually linearly independent. Remarkably, $\mathbf{p}_n$ can be chosen using only knowledge of $\mathbf{p}_{n-1}$, $\nabla f(\mathbf{x}_{n-1})$ and $\nabla f(\mathbf{x}_n)$ (see [69] for proof):

$$\mathbf{p}_n = \nabla f_n + \left(\frac{\nabla f_n^\top \nabla f_n}{\nabla f_{n-1}^\top \nabla f_{n-1}} \right) \mathbf{p}_{n-1}$$

Using conjugate gradients, an N-dimensional form can be minimized in *at most N steps*. Figure A.3 shows the algorithm applied to minimization of the quadratic shown in Figure A.1. The algorithm is run from three different starting points. In each case, the minimum is reached in only two steps.

A geometrical explanation of the power of this method is as follows (illustrated in Figure A.4). Given *any* starting point and *any* descent direction, line minimization ends at a point on one of the contour ellipses tangent to the descent direction. The conjugate point is the corresponding tangent point on the other side of the ellipse, and the line joining the two points is guaranteed to pass through the centre of the

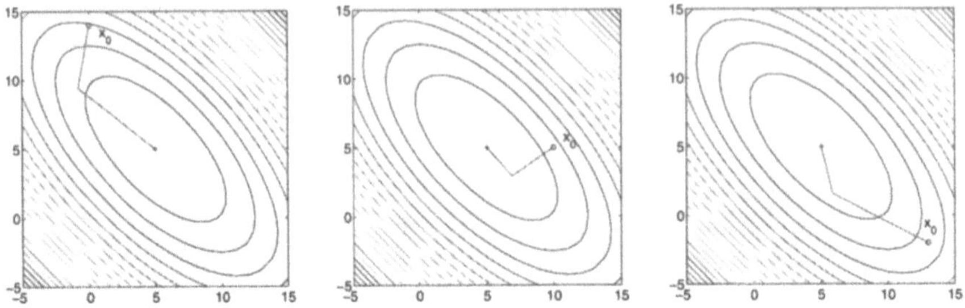

Figure A.3: The method of conjugate gradient descent applied to the minimization of the quadratic function shown in Figure A.1. Regardless of the starting point, the minimum is reached in at most two steps.

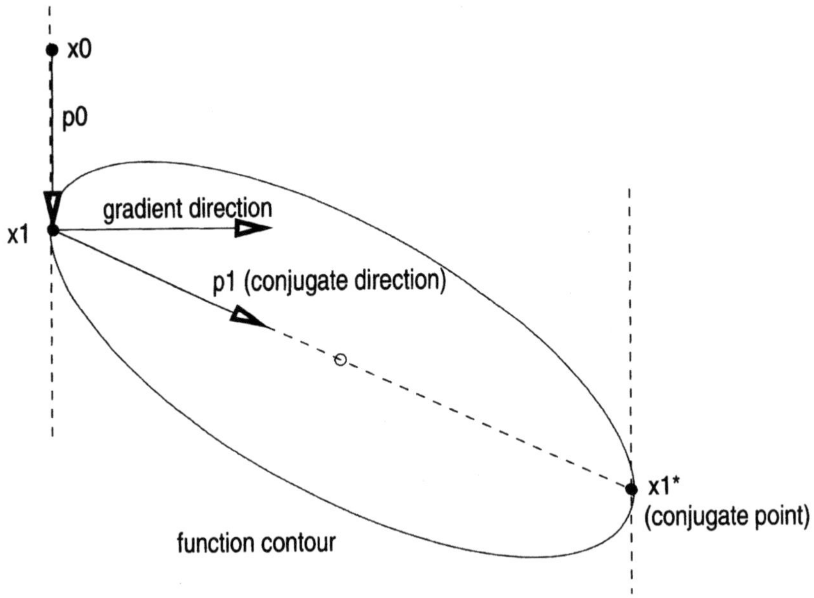

Figure A.4: A geometric interpretation of the conjugate gradient method. After each line minimization, such as at point $\mathbf{x}_1$, the next conjugate direction is the line joining $\mathbf{x}_1$ with its conjugate point $\mathbf{x}_1^*$.

ellipse. The direction of that line, $\mathbf{p}_1$, is defined implicitly by $\mathbf{p}_1^\top \mathbf{C}\mathbf{p}_0 = 0$, as in equation (A.1).

Conjugate gradient descent is far superior to steepest descent, and is the method of choice for the minimization of quadratic functions when the number of parameters N is very large.

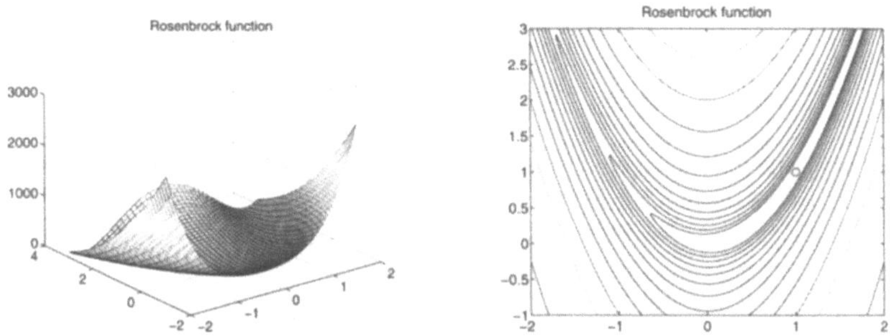

Figure A.5: Rosenbrock's function.

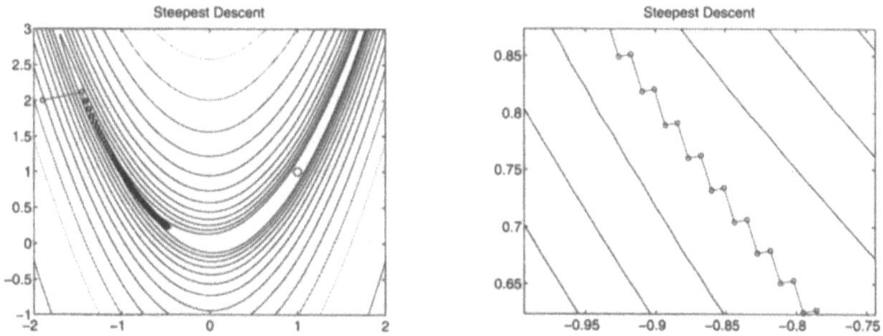

Figure A.6: Steepest descent applied to the Rosenbrock function. Performance is even worse than in the quadratic example. The zig-zag behaviour is clear in the zoomed view (100 iterations).

A.3 General non-linear functions

In the case where $f(\mathbf{x})$ is not a quadratic function of $\mathbf{x}$, the above method cannot be used. A toy example of such a function is Rosenbrock's function of two variables

$$f(x, y) = 100(y - x^2)^2 + (1 - x)^2$$

This function, shown in Figure A.5, has its minimum at $(1, 1)$.

A naive approach to minimizing this function, using the method of steepest descent, meets with even poorer convergence performance than in the quadratic example above. This is demonstrated in Figure A.6. Even after 100 iterations, the iterate is nowhere near the function minimum. Note that, unlike the quadratic case, there is no closed formed solution to the 1D line-minimization sub-problems, so an iterative line-minimization algorithm must be used [68, 116].

A.3.1 Newton methods

The standard approach to minimizing such functions is to use the N-dimensional analog of Newton's method [68, 116]. Approximating $f(\mathbf{x})$ locally by its Taylor series expansion gives

$$f(\mathbf{x}+\mathbf{h}) \approx f(\mathbf{x}) + \nabla f(\mathbf{x})^\top \mathbf{h} + \frac{1}{2}\mathbf{h}^\top \mathtt{H}(\mathbf{x})\mathbf{h}$$

which is a quadratic in terms of the gradient $\nabla f(\mathbf{x})$ and Hessian $\mathtt{H}(x)$ of $f(\mathbf{x})$. The minimum of the local quadratic approximation is at

$$\mathbf{x}_{\min} = \mathbf{x}_n - \mathtt{H}(\mathbf{x}_n)^{-1}\nabla f(\mathbf{x}_n)$$

Naively continuing the analogy with Newton's method would mean minimizing successive local quadratic approximations until the minimum of $f(\mathbf{x})$ is reached. Such a method is appealing since it has a quadratic rate of convergence. However, it suffers from very poor *global convergence* characteristics, meaning that it will only converge if the initial iterate is suitably close to a local minimum of $f(\mathbf{x})$. The reason for this is that the quadratic approximation is generally poor far from the minimum. Furthermore, if $\mathbf{f}(x)$ is non-convex, the quadratic approximation at some points $\mathbf{x}$ will have an *indefinite* Hessian. In this case the quadratic is not bounded below, and hence does not have a unique minimum.

A simple and effective globalization strategy is the "damped" Newton method [68]. In modern terminology, this algorithm is closely related to the "trust region" strategy [24, 49, 68, 102]. It works by augmenting the Hessian $\mathtt{H}(\mathbf{x})$ by adding some multiple of the identity. The Newton step becomes

$$\begin{aligned} \mathbf{x}_{\min} &= \mathbf{x}_n - [\mathtt{H}(\mathbf{x}_n) + \lambda \mathtt{I}]^{-1}\nabla f(\mathbf{x}_n) \\ &= \mathbf{x}_n - \mathtt{H}_{\text{aug}}(\mathbf{x}_n, \lambda)^{-1}\nabla f(\mathbf{x}_n) \end{aligned} \tag{A.2}$$

When λ is small, the method is equivalent to Newton's method, and has the same desirable quadratic convergence rate. When λ is large, the identity dominates and the method tends toward steepest-descent, which is slow, but guarantees a decrease in $f(\mathbf{x})$ even when far from the local minimum. The method is completed by some simple rules for controlling the size of λ. The complete algorithm is as follows.

1. Set $\lambda = 1$.
2. Solve $\boldsymbol{\delta}\mathbf{x} = -\mathtt{H}_{\text{aug}}(\mathbf{x}, \lambda)^{-1}\nabla f(\mathbf{x})$.
3. If $f(\mathbf{x}_n + \boldsymbol{\delta}\mathbf{x}) > f(\mathbf{x}_n)$, increase λ ($\times 10$ say) and go to step 2.
4. Otherwise, decrease λ ($\times 0.1$ say), let $\mathbf{x}_{n+1} = \mathbf{x}_n + \boldsymbol{\delta}\mathbf{x}$, and go to step 2.

The algorithm is run until the function gradient $\nabla f(\mathbf{x})$ or the step-length $\boldsymbol{\delta}\mathbf{x}$ is deemed to be suitably small. Figure A.7 shows the method applied to find the minimum of the Rosenbrock function. A contour on each of the successive quadratic sub-problems is also shown. The algorithm converges after 31 iterations. The damped Newton method has the advantages of being efficient, globally convergent, and very simple to implement.

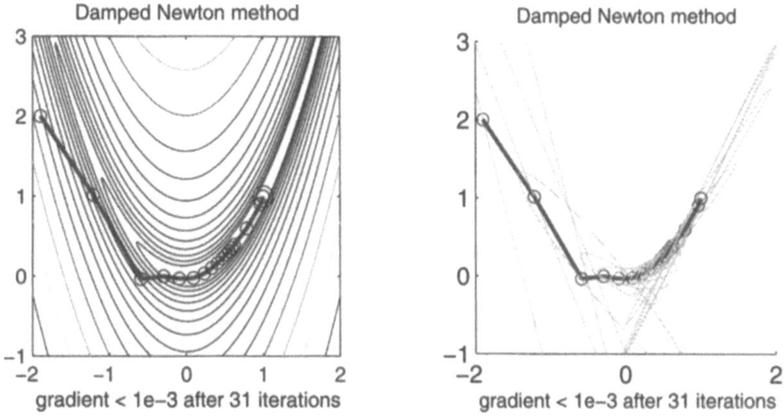

Figure A.7: (Right) Successive iterates in the damped Newton method, applied to find the Rosenbrock minimum. (Left) Contours of the successive quadratic sub-problems. The algorithm converges in 31 iterations.

A.3.2 Computing a Newton step

In the above example, the Newton step $\boldsymbol{\delta}\mathbf{x}$ is trivial to compute by inverting the 2×2 augmented Hessian. When N is extremely large however, direct inversion is infeasible. In this case, $\boldsymbol{\delta}\mathbf{x}$ is computed by iteratively solving

$$\mathtt{H}_{\text{aug}}(\mathbf{x}, \lambda)\boldsymbol{\delta}\mathbf{x} = -\nabla f(\mathbf{x})$$

As stated earlier, conjugate gradient descent is an efficient solver for this problem. Furthermore, the CG termination tolerance may be very "loose". There is nothing to be gained by wasting large amounts of computational effort solving for the Newton step to a high degree of accuracy, particularly since, if the step does not lead to an overall function decrease, it will be discarded and λ modified. For this reason, the CG iteration is terminated when the CG residual $(\mathtt{H}_{\text{aug}}\boldsymbol{\delta}\mathbf{x}_n + \nabla f)$ falls to one tenth of its initial value. This typically requires only a few tens of iterations.

A.4 Non-linear least-squares

It is very common in vision problems for a cost function $f(\mathbf{x})$ to be the sum of a large number of squared residuals :

$$f(\mathbf{x}) = \sum_{i=1}^{M} r_i^2$$

If each residual depends non-linearly on the parameters $\mathbf{x}$ then the minimization of $f(\mathbf{x})$ is a non-linear least squares problem. The $M \times N$ Jacobian of the vector of residuals $\mathbf{r}$ is defined as

$$\mathtt{J}(\mathbf{x}) = \begin{pmatrix} \frac{\partial r_1}{\partial x_1} & \cdots & \frac{\partial r_1}{\partial x_N} \\ \vdots & \ddots & \\ \frac{\partial r_M}{\partial x_1} & & \frac{\partial r_M}{\partial x_N} \end{pmatrix}$$

Hence

$$\nabla f(\mathbf{x}) = 2\mathtt{J}^\top \mathbf{r}$$

$$\mathtt{H}(\mathbf{x}) = 2\mathtt{J}^\top \mathtt{J} + \sum_{i=1}^{M} r_i \frac{\mathrm{d}^2 r_i}{\mathrm{d}\mathbf{x}^2}$$

Note that the second-order term in the Hessian $\mathtt{H}(\mathbf{x})$ is multiplied by the residuals r_i. In most problems, the residuals will typically be small. Also, at the minimum, the residuals will typically be distributed with mean equal to zero. For these reasons, the second-order term is often ignored, leading to the *Gauss–Newton* approximation of the Hessian

$$\mathtt{H}(\mathbf{x}) = 2\mathtt{J}^\top \mathtt{J}$$

Using this approximation, explicit (and potentially expensive) computation of the full Hessian is avoided when solving non-linear least squares problems. Substituting the Gauss–Newton approximation into the damped Newton method of equation (A.2) produces the well-known Levenberg–Marquardt algorithm [68, 95, 99, 116].

References

[1] `http://www.apple.com/quicktime/qtvr.`

[2] `http://www.cognitech.com.`

[3] `http://www.robots.ox.ac.uk/~improofs.`

[4] `http://www.salientstills.com.`

[5] `http://www.smoothmove.com.`

[6] **E. Adelson**. *Layered representations for vision and video.* In ICCV Workshop on the Representation of Visual Scenes, 1995.

[7] **M. Aggarwal and N. Ahuja**. *High dynamic range panoramic imaging.* In Proc. 8th International Conference on Computer Vision, Vancouver, Canada, pages I:2–9, 2001.

[8] **M. Aggarwal, H. Hua, and N. Ahuja**. *On cosine-fourth and vignetting effects in real lenses.* In Proc. 8th International Conference on Computer Vision, Vancouver, Canada, pages I:472–479, 2001.

[9] **P. Anandan and M. Irani**. *Video representation and manipulation using mosaics.* In R. Benosman and S. B. Kang, editors, Panoramic Vision: Sensors, Theory, Applications, pages 393–424. Springer, New York, 2001.

[10] **S. Baker and T. Kanade**. *Limits on super-resolution and how to break them.* In Proc. IEEE Conference on Computer Vision and Pattern Recognition, 2000.

[11] **B. Bascle, A. Blake, and A. Zisserman**. *Motion deblurring and super-resolution from an image sequence.* In Proc. 4th European Conference on Computer Vision, Cambridge, pages 312–320. Springer-Verlag, 1996.

[12] **B. Bascle and R. Deriche**. *Region tracking through image sequences.* In Proc. 5th International Conference on Computer Vision, Boston, 1995.

[13] **P. Beardsley, P. Torr, and A. Zisserman**. *3D model acquisition from extended image sequences.* In Proc. European Conference on Computer Vision, LNCS 1064/1065, pages 683–695. Springer-Verlag, 1996.

[14] **J. Besag**. *Spatial interaction and the statistical analysis of lattice systems.* Journal of the Royal Statistical Society, pages 192–236, 1974.

[15] **J. Besag**. *On the statistical analysis of dirty pictures.* Journal of the Royal Statistical Society, B-48(3):259–302, 1986.

[16] **M. J. Black and P. Anandan**. *The robust estimation of multiple motions: Parametric and piecewise-smooth flow-fields.* Computer Vision and Image Understanding, 63(1):75–104, January 1996.

[17] **A. Blake and A. Zisserman**. *Visual Reconstruction.* MIT Press, Cambridge, USA, August 1987.

[18] **M. Bober, N. Georgis, and J. Kittler**. *On accurate and robust estimation of fundamental matrix.* In Proc. 7th British Machine Vision Conference, Edinburgh, 1996.

[19] **S. Borman, K. Sauer, and C. Bouman**. *Nonlinear prediction methods for estimation of clique weighting parameters in nongaussian image models.* In SPIE Conferences, volume 3459, San Diego, CA, July 1998.

[20] **S. Borman and R. L. Stevenson**. *Simultaneous multi-frame MAP super-resolution video enhancement using spatio-temporal priors.* In Proc. IEEE International Conference on Image Processing, 1999.

[21] **C. Bouman and K. Sauer**. *A generalized gaussian image model for edge-preserving map estimation.* Image Processing, 2(3):296–310, July 1993.

[22] **S. K. Bramble and P. M. Fabrizi**. *Observations on the effects of image processing functions on fingermark data in the fourier domain.* In SPIE Conferences, volume 2567, pages 138–144, 1995.

[23] **S. K. Bramble and J. Jackson**. *Operational experience of fingermark enhancement by frequency domain filtering.* Journal of Forensic Sciences, 39:920–932, 1994.

[24] **R. H. Byrd, R. B. Schnabel, and G. A. Shultz**. *Approximate solution of the trust region problem by minimization over two-dimensional subspaces.* Mathematical Programming, 40:247–263, 1988.

[25] **A. Can, C. V. Stewart, and B. Roysam**. *Robust hierarchical algorithm for constructing a mosaic from images of the curved human retina.* In Proc. IEEE Conference on Computer Vision and Pattern Recognition, Fort Collins, pages II:286–292, 1999.

[26] **F. M. Candocia and J. C. Principe**. *Super-resolution of images based on local correlations.* IEEE Trans. Neural Networks, 10(2):372, 1999.

[27] **J. F. Canny**. *A computational approach to edge detection.* IEEE Transactions on Pattern Analysis and Machine Intelligence, 8(6):679–698, 1986.

[28] **D. P Capel and A. Zisserman**. *Automated mosaicing with super-resolution zoom*. In Proc. IEEE Conference on Computer Vision and Pattern Recognition, Santa Barbara, pages 885–891, June 1998.

[29] **D. P. Capel and A. Zisserman**. *Super-resolution enhancement of text image sequences*. In Proc. International Conference on Pattern Recognition, pages 600–605, 2000.

[30] **D. P. Capel, A. Zisserman, S. Bramble, and D. Compton**. *An automatic method for the removal of unwanted, non-periodic patterns from forensic images*. In Proc. of SPIE, Boston, MA, volume 3576, 1-6 November 1998.

[31] **R. Chan, T. Chan, and C. Wong**. *Cosine transform based preconditioners for total variation minimization problems in image processing*. In Proc. 2nd IMACS Symposium on Iterative Methods, Blagoevgrad, Bulgaria, pages 311–329, June 1995.

[32] **T. Chan, P. Blomgren, P. Mulet, and C. K. Wong**. *Total variation image restoration: Numerical methods and extensions*. In Proc. IEEE International Conference on Image Processing, page III:384, 1997.

[33] **T. F. Chan, G. H. Golub, and P. Mulet**. *A nonlinear primal-dual method for total variation-based image restoration*. SIAM Journal on Scientific Computing, 20(6):1964–1977, 1999.

[34] **P. Cheeseman, B. Kanefsky, R. Kraft, and J. Stutz**. *Super-resolved surface reconstruction from multiple images*. Technical report, NASA, 1994.

[35] **S. Chen**. *Quicktime VR – an image-based approach to virtual environment navigation*. In Proceedings of the ACM SIGGRAPH Conference on Computer Graphics, 1995.

[36] **S. Chen and L. Williams**. *View interpolation for image synthesis*. In Proceedings of the ACM SIGGRAPH Conference on Computer Graphics, 1993.

[37] **S. Coorg, N. Master, and S. Teller**. *Acquisition of a large pose-mosaic dataset*. In Proc. IEEE Conference on Computer Vision and Pattern Recognition, Santa Barbara, pages 872–878, 1998.

[38] **S. Coorg and S. Teller**. *Spherical mosaics with quaternions and dense correlation*. International Journal of Computer Vision, 37(3):259–273, June 2000.

[39] **A. Criminisi, I. Reid, and A. Zisserman**. *RaDiCo – radial distortion correction from single view*. Technical Report OUEL, Dept. Engineering Science, University of Oxford, 1998.

[40] **A. Criminisi, I. Reid, and A. Zisserman**. *Single view metrology*. International Journal of Computer Vision, 40(2):123–148, November 2000.

[41] **G. Cross**. *Surface Reconstruction from Image Sequences: Texture and Apparent Contour Constraints*. PhD thesis, University of Oxford, 2000.

[42] **G. Cross, A. W. Fitzgibbon, and A. Zisserman**. *Parallax geometry of smooth surfaces in multiple views*. In Proc. 7th International Conference on Computer Vision, Kerkyra, Greece, pages 323–329, September 1999.

[43] **G. Cross and A. Zisserman**. *Quadric surface reconstruction from dual-space geometry*. In Proc. 6th International Conference on Computer Vision, Bombay, India, pages 25–31, January 1998.

[44] **G. Cross and A. Zisserman**. *Surface reconstruction from multiple views using apparent contours and surface texture*. In A. Leonardis, F. Solina, and R. Bajcsy, editors, NATO Advanced Research Workshop on Confluence of Computer Vision and Computer Graphics, Ljubljana, Slovenia, pages 25–47, 2000.

[45] **J. E. Davis**. *Mosaics of scenes with moving objects*. In Proc. IEEE Conference on Computer Vision and Pattern Recognition, Santa Barbara, pages 354–360, 1998.

[46] **L. de Agapito, E. Hayman, and I. Reid**. *Self-calibration of a rotating camera with varying intrinsic parameters*. In Proc. 9th British Machine Vision Conference, Southampton, 1998.

[47] **F. Dellaert, S. Thrun, and C. Thorpe**. *Jacobian images of super-resolved texture maps for model based motion estimation and tracking*. In Proc. IEEE Workshop on Applications of Computer Vision, 1998.

[48] **F. Dellaert, S. Thrun, and C. Thorpe**. *Mosaicing a large number of widely dispersed, noisy, and distorted images: A Bayesian approach*. Technical report, School of Computer Science, Carnegie Mellon University, 1999.

[49] **J. E. Dennis Jr. and R. N. Schnabel**. *Numerical methods for unconstrained optimization and nonlinear equations*, volume 16. Society for Industrial and Applied Mathematics (SIAM), Philadelphia, PA, 1996.

[50] **F. Devernay and O. D. Faugeras**. *Automatic calibration and removal of distortion from scenes of structured environments*. In SPIE Conferences, volume 2567, San Diego, CA, July 1995.

[51] **A. Efros and T. Leung**. *Texture synthesis by non-parametric sampling*. In Proc. 7th International Conference on Computer Vision, Kerkyra, Greece, pages 1039–1046, September 1999.

[52] **M. Elad**. *Super-resolution reconstruction of images*. PhD thesis, Technion, Israel, 1996.

[53] **M. Elad and A. Feuer**. *Super-resolution reconstruction of continuous image sequence - the non-casual approach*. Technical report, Technion, Israel, 1996.

[54] **M. Elad and A. Feuer**. *Super-resolution reconstruction of continuous image sequences*. In Proc. IEEE International Conference on Image Processing, page 27AP5, 1999.

[55] **M. Elad and A. Feuer**. *Super-resolution reconstruction of image sequences.* IEEE Transactions on Pattern Analysis and Machine Intelligence, 21(9):817, September 1999.

[56] **M. Elad and A. Feuer**. *Superresolution restoration of an image sequence: Adaptive filtering approach.* Image Processing, 8(3):387, March 1999.

[57] **H. Engl, M. Hanke, and A. Neubauer**. *Regularization of Inverse Problems.* Kluwer Academic Publishers, Dordrecht, 1996.

[58] **P. E. Eren, M. I. Sezan, and A. M. Tekalp**. *Robust region-based high-resolution image reconstruction from low-resolution video.* In Proc. IEEE International Conference on Image Processing, page 16P7, 1996.

[59] **P. E. Eren, M. I. Sezan, and A. M. Tekalp**. *Robust, object based high resolution image reconstruction from low resolution video.* IP, 6(10):1446–1451, October 1997.

[60] **M. A. Fischler and R. C. Bolles**. *Random sample consensus: A paradigm for model fitting with applications to image analysis and automated cartography.* Comm. Assoc. Comp. Mach., 24(6):381–395, 1981.

[61] **W. Freeman and E. Pasztor**. *Learning low-level vision.* In Proc. International Conference on Computer Vision, pages 1182–1189, 1999.

[62] **W. Freeman and E. Pasztor**. *Learning to estimate scenes from images.* Technical report, Mitsubishi Electric Research Laboratory, 1999.

[63] **W. Freeman and E. Pasztor**. *Markov networks for low-level vision.* Technical report, Mitsubishi Electric Research Laboratory, 1999.

[64] **W. T. Freeman, E. C. Pasztor, and O. T. Carmichael**. *Learning low-level vision.* International Journal of Computer Vision, 40(1):25–47, October 2000.

[65] **S. Geman and D. Geman**. *Stochastic relaxation, gibbs distributions, and the bayesian restoration of images.* IEEE Transactions on Pattern Analysis and Machine Intelligence, 6(6):721–741, November 1984.

[66] **A. R. German**. *Analog & Digital Image Processing of Latent Fingerprints (Eds. H.C. Lee and R.E. Gaensslen)*, chapter 7, pages 193–208. Elsevier, 1991.

[67] **P. E. Gill and W. Murray**. *Algorithms for the solution of the nonlinear least-squares problem.* SIAM J. Numerical Analysis, 15(5):977–992, 1978.

[68] **P. E. Gill, W. Murray, and M. H. Wright**. *Practical Optimization.* Academic Press, London, 1981.

[69] **G. H. Golub and C. F. Van Loan**. *Matrix Computations.* The Johns Hopkins University Press, Baltimore, MD, second edition, 1989.

[70] **R. Gonzalez and P. Wintz**. *Image Processing.* Addison-Wesley, 1987.

[71] **C. Groetsch**. *The Theory of Tikhonov Regularization for Fredholm Equations of the First Kind.* Pitman, 1984.

[72] **B. Gunturk, Y. Altunbasak, and R. Mersereau**. *Bayesian resolution-enhancement framework for transform-coded video.* In Proc. IEEE International Conference on Image Processing, 2001.

[73] **C. J. Harris and M. Stephens**. *A combined corner and edge detector.* In Proc. 4th Alvey Vision Conference, Manchester, pages 147–151, 1988.

[74] **R. I. Hartley**. *Self-calibration from multiple views with a rotating camera.* In Proc. European Conference on Computer Vision, LNCS 800/801, pages 471–478. Springer-Verlag, 1994.

[75] **R. I. Hartley**. *Self-calibration of stationary cameras.* International Journal of Computer Vision, 22(1):5–23, February 1997.

[76] **R. I. Hartley and A. Zisserman**. *Multiple View Geometry in Computer Vision.* Cambridge University Press, 2000.

[77] **P. J. Huber**. *Robust Statistics.* John Wiley and Sons, 1981.

[78] **P. J. Huber**. *Projection pursuit.* Annals of Statistics, 13:433–475, 1985.

[79] **M. Irani and P. Anandan**. *Video indexing based on mosaic representations.* Proceedings of IEEE, 86(5):905–921, May 1998.

[80] **M. Irani and P. Anandan**. *About direct methods.* In W. Triggs, A. Zisserman, and R. Szeliski, editors, Vision Algorithms: Theory and Practice, LNCS. Springer-Verlag, 2000.

[81] **M. Irani, P. Anandan, J. Bergen, R. Kumar, and S. Hsu**. *Efficient representations of video sequences and their applications.* Signal Processing: Image Communication, 8(4):327–351, May 1996.

[82] **M. Irani, P. Anandan, and S. Hsu**. *Mosaic based representations of video sequences and their applications.* In Proc. 5th International Conference on Computer Vision, Boston, pages 605–611, 1995.

[83] **M. Irani, S. Hsu, and P. Anandan**. *Video compression using mosaic representations.* Signal Processing: Image Communication, 7(4):529–552, November 1995.

[84] **M. Irani and S. Peleg**. *Improving resolution by image registration.* Graphical Models and Image Processing, 53:231–239, 1991.

[85] **M. Irani and S. Peleg**. *Motion analysis for image enhancement:resolution, occlusion, and transparency.* Journal of Visual Communication and Image Representation, 4:324–335, 1993.

[86] **M. Irani, B. Rousso, and S. Peleg**. *Detecting and tracking multiple moving objects using temporal integration*. In G. Sandini, editor, Proc. 2nd European Conference on Computer Vision, Santa Margharita Ligure, Italy, pages 282–287. Springer-Verlag, 1992.

[87] **M. Irani, B. Rousso, and S. Peleg**. *Computing occluding and transparent motions*. International Journal of Computer Vision, 12(1):5–16, 1994.

[88] **A. K. Jain**. *Fundamentals of Digital Image Processing*. Prentice-Hall, 1989.

[89] **S. B. Kang and R. Szeliski**. *3-d scene data recovery using omnidirectional multibaseline stereo*. In Proc. IEEE Conference on Computer Vision and Pattern Recognition, pages 364–370, 1996.

[90] **S. B. Kang and R. Szeliski**. *3-d scene data recovery using omnidirectional multibaseline stereo*. International Journal of Computer Vision, 25(2):167–183, November 1997.

[91] **M. Kendall and A. Stuart**. *The Advanced Theory of Statistics*. Charles Griffin and Company, London, 1983.

[92] **D. Keren, S. Peleg, and R. Brada**. *Image sequence enhancement using subpixel displacements*. In Proc. IEEE Conference on Computer Vision and Pattern Recognition, pages 742–746, 1988.

[93] **R. Kumar, P. Anandan, M. Irani, J. Bergen, and K. Hanna**. *Representation of scenes from collections of images*. In ICCV Workshop on the Representation of Visual Scenes, 1995.

[94] **H. J. Larson**. *Introduction to Probability Theory and Statistical Inference*. John Wiley, 1982.

[95] **K. Levenberg**. *A method for the solution of certain non-linear problems in least squares*. Quarterly Applied Mathematics, II(2):164–168, 1944.

[96] **S. Z. Li**. *Markov Random Field Modeling in Computer Vision*. Springer-Verlag, Tokyo, 1995.

[97] **S. Mann and R. W. Picard**. *Virtual bellows: Constructing high quality stills from video*. In Proc. IEEE International Conference on Image Processing, 1994.

[98] **S. Mann and R. W. Picard**. *Video orbits of the projective group: A new perspective on image mosaicing*. Technical report, MIT, 1996.

[99] **D. W. Marquardt**. *An algorithm for least-squares estimation of nonlinear parameters*. Journal of the Society for Industrial and Applied Mathematics, 11(2):431–441, 1963.

[100] **L. McMillan and G. Bishop**. *Plenoptic modeling: An image-based rendering system*. In Proceedings of the ACM SIGGRAPH Conference on Computer Graphics, 1995.

[101] **J. More and G. Toraldo**. *On the solution of large quadratic programming problems with bound constraints*. SIAM J. Optimization, 1(1):93–113, 1991.

[102] **J. J Moré and D. C. Sorensen**. *Computing a trust region step*. SIAM Journal on Scientific and Statistical Computing, 4(3):553–572, 1983.

[103] **N. Nguyen, P. Milanfar, and G. H. Golub**. *A computationally efficient image superresolution algorithm*. Image Processing, March 2001.

[104] **B. A. Olshausen and D. J. Field**. *Natural image statistics and efficient coding*. Network, pages 7:333–339, 1996.

[105] **A. J. Patti and Y. Altunbasak**. *Artifact reduction for set theoretic super resolution image reconstruction with edge adaptive constraints and higher-order interpolants*. Image Processing, 10(1):179–186, January 2001.

[106] **A. J. Patti, M. I. Sezan, and A. M. Tekalp**. *Robust methods for high-quality stills from interlaced video in the presence of dominant motion*. IEEE Trans. Circuits and Systems for Video Technology, 7(2):328–342, April 1997.

[107] **A. J. Patti, M. I. Sezan, and A. M. Tekalp**. *Superresolution video reconstruction with arbitrary sampling lattices and nonzero aperture time*. Image Processing, 6(8):1064–1076, August 1997.

[108] **S. Peleg**. *Panoramic mosaics by manifold projection*. Technical report, Hebrew University of Jerusalem, 1997.

[109] **S. Peleg, M. Ben-Ezra, and Y. Pritch**. *Omnistereo: Panoramic stereo imaging*. IEEE Transactions on Pattern Analysis and Machine Intelligence, 23(3):279–290, March 2001.

[110] **S. Peleg, M. Ben-Ezra, and Y. Pritch**. *Panoramic imaging with horizontal stereo*. In R. Benosman and S. B. Kang, editors, Panoramic Vision: Sensors, Theory, Applications, pages 143–160, 2001.

[111] **S. Peleg and J. Herman**. *Panoramic mosaics by manifold projection*. In Proc. IEEE Conference on Computer Vision and Pattern Recognition, 1997.

[112] **S. Peleg and J. Herman**. *Panoramic mosaics with videobrush*. In Image Understanding Workshop (DARPA), pages 261–264, 1997.

[113] **S. Peleg, D. Keren, and L. Schweitzer**. *Improving image resolution using subpixel motion*. Pattern Recognition Letters, 5:223–226, 1987.

[114] **S. Peleg, B. Rousso, A. Rav-Acha, and A. Zomet**. *Mosaicing with strips on adaptive manifolds*. In Panoramic Vision: Sensors, Theory, Applications, pages 309–325, 2001.

[115] **W. K. Pratt**. *Digital Image Processing*. John Wiley & Sons, 1991.

[116] **W. Press, B. Flannery, S. Teukolsky, and W. Vetterling**. *Numerical Recipes in C*. Cambridge University Press, 1988.

[117] **S. E. Reichenbach, S. K. Park, and R. Narayanswamy**. *Characterizing digitial image aquisition devices.* Optical Engineering, 30(2):170–177, 1991.

[118] **J. A. Rice**. *Mathematical Statistics and Data Analysis*. Wadsworth and Brooks, California, 1988.

[119] **M. A. Robertson, S. Borman, and R. L. Stevenson**. *Estimation theoretic approach to dynamic range improvement through multiple exposures*. In Proc. IEEE International Conference on Image Processing, page 27AO3, 1999.

[120] **B. Rousso, S. Peleg, and I. Finci**. *Mosaicking with generalized strips*. In Image Understanding Workshop (DARPA), pages 255–260, 1997.

[121] **B. Rousso, S. Peleg, I. Finci, and A. Rav-Acha**. *Universal mosaicing using pipe projection*. In Proc. 6th International Conference on Computer Vision, Bombay, India, pages 945–952, 1998.

[122] **L. Rudin, F. Guichard, and P. Yu**. *Video super-resolution via contrast-invariant motion segmentation and frame fusion (with applications to forensic video evidence)*. In Proc. IEEE International Conference on Image Processing, page 27PS1, 1999.

[123] **P. D. Sampson**. *Fitting conic sections to 'very scattered' data: An iterative refinement of the bookstein algorithm*. Computer Vision, Graphics, and Image Processing, 18:97–108, 1982.

[124] **S. S. Saquib, C. A. Bouman, and K. Sauer**. *ML parameter estimation for markov random fields with applications to bayesian tomography*. Image Processing, 7(7):1029–1044, July 1998.

[125] **S. S. Saquib, J. Zheng, C. A. Bouman, and K. Sauer**. *Provably convergent coordinate descent in statistical tomographic reconstruction*. In Proc. IEEE International Conference on Image Processing, volume 2, 1996.

[126] **H. S. Sawhney, S. Hsu, and R. Kumar**. *Robust video mosaicing through topology inference and local to global alignment*. In Proc. European Conference on Computer Vision, 1998.

[127] **H. S. Sawhney and R. Kumar**. *True multi-image alignment and its application to mosaicing and lens distortion correction*. IEEE Transactions on Pattern Analysis and Machine Intelligence, 21(3):235–243, 1999.

[128] **H. S. Sawhney, R. Kumar, G. Gendel, J. Bergen, D. Dixon, and V. Paragano**. *Videobrush: Experiences with consumer video mosaicing*. In Proc. IEEE Workshop on Applications of Computer Vision, 1998.

[129] **Y. Y. Schechner and S. K. Nayar**. *Generalized mosaicing.* In Proc. 8th International Conference on Computer Vision, Vancouver, Canada, pages I:17–24, 2001.

[130] **C. Schmid, R. Mohr, and C. Bauckhage**. *Comparing and evaluating interest points.* In Proc. International Conference on Computer Vision, pages 230–235, 1998.

[131] **R. R. Schultz and R. L. Stevenson**. *Extraction of high-resolution frames from video sequences.* IEEE Transactions on Image Processing, 5(6):996–1011, June 1996.

[132] **K. Schutte and A. Vossepoel**. *Accurate mosaicing of scanned maps,or how to generate a virtual A0 scanner.* In Annual Conference for the Advance School of Computing, pages 353–359, 1995.

[133] **C. Segall, R. Molina, A. Katsaggelos, and J. Mateos**. *Bayesian high-resolution reconstruction of low-resolution compressed video.* In Proc. IEEE International Conference on Image Processing, 2001.

[134] **S. Seitz and C. Dyer**. *Physically-valid view synthesis by image interpolation.* In ICCV Workshop on the Representation of Visual Scenes, 1995.

[135] **A. Shashua and S. Toelg**. *The quadric reference surface: Theory and applications.* International Journal of Computer Vision, 23(2):185–198, 1997.

[136] **H. Shekarforoush and R. Challappa**. *Data-driven multichannel super-resolution with application to video sequences.* Journal of the Optical Society of America, 16(3):481–492, March 1999.

[137] **H. Shekarforoush and R. Chellappa**. *Adaptive super-resolution for predator video sequences.* In Image Understanding Workshop (DARPA), pages 995–1002, 1998.

[138] **S. D. Silvey**. *Statistical Inference.* Penguin, Harmondsworth, Middlesex, 1970.

[139] **E. P. Simoncelli**. *Statistical models for images: Compression, restoration and synthesis.* In Asilomar97, 1997.

[140] **C. Slama**. *Manual of Photogrammetry.* American Society of Photogrammetry, Falls Church, VA, 4th edition, 1980.

[141] **V. N. Smelyanskiy, P. Cheeseman, D. Maluf, and R. Morris**. *Bayesian super-resolved surface reconstruction from images.* In Proc. IEEE Conference on Computer Vision and Pattern Recognition, Hilton Head Island, pages I:375–382, 2000.

[142] **M. R. Spiegel**. *Theory and Problems of Statistics, 2nd edition.* McGraw Hill, 1992.

[143] **G. P. Stein**. *Lens distortion calibration using point correspondences*. In Proc. IEEE Conference on Computer Vision and Pattern Recognition, Puerto Rico, 1997.

[144] **D. Strong, D. Blomgren, and T. Chan**. *Spatially adaptive local feature-driven total variation minimizing image restoration*. In SPIE Conferences, volume 3167, 1997.

[145] **R. Szeliski**. *Image mosaicing for tele-reality applications*. Technical report, Digital Equipment Corporation, Cambridge, USA, 1994.

[146] **R. Szeliski and S. Heung-Yeung**. *Creating full view panoramic image mosaics and environment maps*. In Proceedings of the ACM SIGGRAPH Conference on Computer Graphics, 1997.

[147] **R. Szeliski and S. B. Kang**. *Direct methods for visual scene reconstruction*. In ICCV Workshop on the Representation of Visual Scenes, 1995.

[148] **S. Teller**. *Automated urban model acquisition: Project rationale and status*. In Image Understanding Workshop (DARPA), pages 455–462, 1998.

[149] **A. N. Tikhonov and V. Y. Arsenin**. *Solutions of Ill-Posed Problems*. V.H. Winston & Sons, John Wiley & Sons, Washington DC, 1977.

[150] **N. Tiller and T. Tiller**. *Conviction through enhanced fingerprint identification*. FBI Law Enforcement Bulletin, 61(12):16–17, December 1992.

[151] **B. Tom and A. Katsaggelos**. *Reconstruction of a high-resolution image by simultaneous registration, restoration and interpolation of low-resolution images*. In Proc. IEEE International Conference on Image Processing, 1995.

[152] **B. C. Tom and A. K. Katsaggelos**. *Resolution enhancement of video sequences using motion compensation*. In Proc. IEEE International Conference on Image Processing, 1996.

[153] **P. H. S. Torr**. *Motion segmentation and outlier detection*. PhD thesis, Dept. of Engineering Science, University of Oxford, 1995.

[154] **P. H. S. Torr and D. W. Murray**. *Outlier detection and motion segmentation*. Technical Report 1987/93, University of Oxford, 1993.

[155] **P. H. S. Torr and D. W. Murray**. *A review of robust methods to estimate the fundamental matrix*. Accepted to IJCV, 1996.

[156] **P. H. S. Torr and A. Zisserman**. *Computing multiple view relations*. OUEL Report, 1997.

[157] **P. H. S. Torr and A. Zisserman**. *Robust parameterization and computation of the trifocal tensor*. Image and Vision Computing, 15:591–605, 1997.

[158] **P. H. S. Torr and A. Zisserman**. *Robust computation and parameterization of multiple view relations.* In Proc. 6th International Conference on Computer Vision, Bombay, India, pages 727–732, January 1998.

[159] **P. H. S. Torr and A. Zisserman**. *MLESAC: A new robust estimator with application to estimating image geometry.* Computer Vision and Image Understanding, 78:138–156, 2000.

[160] **P. H. S. Torr, A. Zisserman, and S. Maybank**. *Robust detection of degenerate configurations for the fundamental matrix.* Computer Vision and Image Understanding, 71(3):312–333, September 1998.

[161] **W. Triggs, P. McLauchlan, R. Hartley, and A. Fitzgibbon**. *Bundle adjustment: A modern synthesis.* In W. Triggs, A. Zisserman, and R. Szeliski, editors, Vision Algorithms: Theory and Practice, LNCS. Springer-Verlag, 2000.

[162] **C. J. Tsai, P. Karunaratne, N. P. Galatsanos, and A. K. Katsaggelos**. *Compressed video enhancement with information from the encoder.* In Proc. IEEE International Conference on Image Processing, 1999.

[163] **R. Tsai and T. Huang**. *Multiframe image restoration and registration.* Advances in Computer Vision and Image Processing, 1:317–339, 1984.

[164] **Y. R. Tsai**. *An efficient and accurate camera calibration technique for 3D machine vision.* In Proc. IEEE Conference on Computer Vision and Pattern Recognition, 1986.

[165] **Y. Tsin, V. Ramesh, and T. Kanade**. *Statistical calibration of the CCD imaging process.* In Proc. 8th International Conference on Computer Vision, Vancouver, Canada, pages I: 480–487, 2001.

[166] **M. Turk and A. P. Pentland**. *Face recognition using eigenfaces.* In Proc. IEEE Conference on Computer Vision and Pattern Recognition, pages 586–591, 1991.

[167] **H. Ur and D. Gross**. *Improved resolution from subpixel shifted pictures.* Graphical Models and Image Processing, 54(2):181–186, March 1992.

[168] **C. Vogel**. *Total variation regularization for ill-posed problems.* Technical report, Department of Mathematical Sciences, Montana State University., 1993.

[169] **C. R. Vogel and M. E. Oman**. *Fast, robust total variation based reconstruction of noisy, blurred images.* Image Processing, 7(6):813–824, June 1998.

[170] **W. J. Watling**. *Using the FFT in forensic digital image enhancement.* Journal of Forensic Identification, 43(6):573–584, 1993.

[171] **J. S. Yedidia, W. T. Freeman, and Y. Weiss**. *Generalized belief propagation.* Technical Report TR-2000-26, Mitsubishi Electric Research Laboratory, 2000.

[172] **L. Zelnik-Manor and M. Irani**. *Multi-frame alignment of planes.* In Proc. IEEE Conference on Computer Vision and Pattern Recognition, Fort Collins, pages I:151–156, 1999.

[173] **L. Zelnik-Manor and M. Irani**. *Multi-frame estimation of planar motion.* IEEE Transactions on Pattern Analysis and Machine Intelligence, 22(10):1105–1116, October 2000.

[174] **J. Zheng, S. S. Saquib, K. Sauer, and C. A. Bouman**. *Parallelizable Bayesian tomography algorithms with rapid, guaranteed convergence.* Image Processing, 9(10):1745–1759, October 2000.

[175] **A. Zomet and S. Peleg**. *Efficient super-resolution and applications to mosaics.* In Proc. International Conference on Pattern Recognition, pages I:579–583, 2000.

Index

MIX
Papier aus verantwortungsvollen Quellen
Paper from responsible sources
FSC® C105338

If you have any concerns about our products,
you can contact us on
ProductSafety@springernature.com

In case Publisher is established outside the EU,
the EU authorized representative is:
Springer Nature Customer Service Center GmbH
Europaplatz 3, 69115 Heidelberg, Germany

Printed by Libri Plureos GmbH
in Hamburg, Germany